AF508677

Advances in Processing and Mechanical Behavior in Lightweight Metals and Alloys

Advances in Processing and Mechanical Behavior in Lightweight Metals and Alloys

Editor

Claudio Testani

MDPI • Basel • Beijing • Wuhan • Barcelona • Belgrade • Manchester • Tokyo • Cluj • Tianjin

Editor
Claudio Testani
CALEF-ENEA CR Casaccia
Italy

Editorial Office
MDPI
St. Alban-Anlage 66
4052 Basel, Switzerland

This is a reprint of articles from the Special Issue published online in the open access journal *Metals* (ISSN 2075-4701) (available at: https://www.mdpi.com/journal/metals/special_issues/processing_mechanical_lightweight_alloys).

For citation purposes, cite each article independently as indicated on the article page online and as indicated below:

LastName, A.A.; LastName, B.B.; LastName, C.C. Article Title. *Journal Name* **Year**, *Volume Number*, Page Range.

ISBN 978-3-0365-2273-9 (Hbk)
ISBN 978-3-0365-2274-6 (PDF)

Contents

About the Editor

Claudio Testani graduate in Aerospace Structural Engineering in 1987 , University of Rome , La Sapienza, then in 2006 has defended the PHD thesis in Material Engineering, Università di Roma "Tor-Vergata". Has a long standing experience in teaching, he his actually Member of the Teaching Board of the PhD school of TorVergata University Rome, Italy. He is a qualified Engineer and has worked for about 30 years in the research centres for advanced structural materials development, applied metallurgy, industrial research and management with 8 patents, more than 100 scientific publications. He has been the R&D Manager for an italian industry before coming back to the research activities. Actually is the technical director of CALEF an Italian R&D research Consortium. He has received the Award: Top Peer Reviewer 2019 for "Material Science" and the Award: Top Peer Reviewer 2019 for "Cross-Field" on Publons global reviewer database.

Preface to "Advances in Processing and Mechanical Behavior in Lightweight Metals and Alloys"

This book is a compilation of recent articles presenting recent developments in synthesis, characterisation, joining and processing of lightweight alloys of the three: Aluminium, Magnesium and Titanium based families including the intermetallic-reinforced matrices.

The subjects are multidisciplinary including: joining processes, friction stir welding (FSW) and Electron Beam Welding, (EBW), thermomechanical processes such us severe plastic deformation (SPD) by Equal Channel Angular Pressing (ECAP) of oxide dispersed Mg alloy and thermal treatments.

This book can provide a source for data, process information on emerging processes, based on the powder metallurgy, that have been proposed and discussed with interesting results on complex and critical joining processes.

Several processes solutions are proposed for the emerging additive manufacturing joining-process and has been intended to provide a room for sharing new ideas, new research results for the various aspects of light alloy processing and characterisation.

Some of the results can be a real base for an effective application in the electronic and industrial application, but obviously there are still many challenges to overcome.

As a Guest Editor, of this Special Issue I really hope that all scientific results in this Special Issue can contribute to the advancement of light weight metal future research and find application in real industrial environment and market.

I would like to warmly thank the authors of the research articles in this Special Issue for their contributions, and all of the reviewers for their efforts in ensuring high-quality publications.

Finally, thanks to the editors of Metals and to the Metals editorial assistants for their valuable and continuous support during the preparation of this volume.

Claudio Testani
Editor

metals

Editorial

Advances in Processing and Mechanical Behavior in Lightweight Metals and Alloys

Claudio Testani

CALEF, Management Department, c/o ENEA CR Casaccia, Via Anguillarese 301, Santa Maria di Galeria, 00123 Rome, Italy; claudio.testani@consorziocalef.it; Tel.: +39-06-3048-4653

1. Introduction and Scope

The demand for lightweight metals and related alloys is still the most suitable solution to many high-tech applications, including sports equipment and automotive components where alternate movements require low inertia. In this Issue, the term light alloy is focused on materials based on aluminum, titanium, and magnesium systems, including the intermetallic-reinforced matrices, and papers dealing with all of the three families have been published. Nevertheless, the processing of light alloys has always been faced with low-formability, narrow-thermal processing windows and even with corrosion issues with respect to the steel family; this is the reason for new processes in terms of both compositions and thermo-mechanical approaches. Emerging processes based on the powder metallurgy have been proposed and discussed, and interesting results have been obtained on complex and critical joining processes, which still stand and are also proposed for the emerging additive manufacturing process.

This Special Issue covers a wide scope in the research field of lightweight metals and alloys and has been intended to provide a space for sharing new ideas, and new research results for the various aspects of light alloy processing and characterisation.

2. Contributions

Eight research articles have been published in this Special Issue of *Metals* and three lightweight metal families, aluminium, titanium and magnesium have been impacted. The subjects are multidisciplinary, including joining processes, friction stir welding (FSW) and Electron Beam Welding, (EBW), thermomechanical processes such as severe plastic deformation (SPD) by Equal Channel Angular Pressing (ECAP) of oxide dispersed Mg alloy.

Moreover, the Special Issue reports interesting improvements obtained by means of innovative thermal treatments, and last but not least, interesting data obtained from the characterisation of ODS nanostructured powder processing. The study of nanostructured metals is very interesting for broader and larger systems, because of the potential improvements in terms of mechanical performance [1], and the current proposed development of a model, by means of convolutional neural networks (CNNs) [2] for the prediction of the secondary dendrite arm spacing (SDAS), which has an industrially acceptable prediction accuracy for the aluminium alloys, is an important step towards a more efficient manufacturing industry.

Two papers investigate the dissimilar joints between Ti-TiB and $(\alpha+\beta)$Ti Alloy.

Loboda et al. [2] reported the structural features of Ti-TiB and $(\alpha+\beta)$ Ti alloys before and after the electron beam weldment process, showing that it is possible to obtain sound joints, even if it is critical to control the TiB fibre orientation in order to avoid the transversal orientation of TiB-reinforcing fibres that can facilitate brittle fracturing.

Mortello et Al. [3] succeeded in obtaining by FSW hybrid AA5083 H111 aluminium alloy and S355J2 grade DH36 structural steel joints. These results are particularly interesting because of the industrial impact in mechanical structures and, as the authors pointed out

Citation: Testani, C. Advances in Processing and Mechanical Behavior in Lightweight Metals and Alloys. *Metals* **2021**, *11*, 1555. https://doi.org/10.3390/met11101555

Received: 23 September 2021
Accepted: 25 September 2021
Published: 29 September 2021

Publisher's Note: MDPI stays neutral with regard to jurisdictional claims in published maps and institutional affiliations.

which: "… enables the ability to design and fabricate components whose properties are customised to locally varying environmental conditions." [3].

Another paper from Nikolić et al. [4] proposed and adopted an automatic inspection model using a computer vision for secondary dendrite arm spacing (SDAS) in aluminium alloys, which represent an example of a Deep Learning (DL) model that has industrially acceptable prediction accuracy: this approach could be the basis for a next generation industrial heavy-metallurgy digitalisation.

Another paper on the development and characterisation of the Strengthening Mechanism and Influence of Hybrid Reinforcements (β-SiCp, Bi and Sb) on the Mechanical Properties of the AZ91 Magnesium Matrix has been proposed by Huang et al. [5]. It was proven that, during the Mg-alloy AZ91 melt–stir casting, a hybrid nano-reinforcements (β-SiCp) and micro-reinforcements (Bi and Sb) composites network was produced. These reinforcements are the base for further diffusion phenomena that result in the creation of Mg2Si (cubic) and SiC (rhombohedral axes), enhancing the microhardness by more than 18% in a 0.5 wt.% SiCp/AZ91 matrix.

The authors [6] studied the nanoprecipitation on the AA7050 aerospace high strength aluminium alloy, after an innovative thermal treatment, and it was proven to be able to increase the toughness in KIC laboratory testing.

The ECAP process is a useful process for metallurgical studies of light alloys and especially for Mg, as shown by Ou et al. [7], with the paper dealing with the Mg97Zn1Y2 alloy powders reinforced by Y_2O_3 particles prepared by simultaneous synthesising of Y_2O_3 particles and compacting of mechanically alloyed powders using equal channel angular pressing. They found that, after mechanical alloying for 20 h, the material consisted of α-Mg, Y and Y2O3 phases. The ECAP-compacted bulk alloy contained the α-Mg matrix and uniformly dispersed Y_2O_3 and MgO phases. A very impressive hardness value was reached (after ECAP route Bc for four passes) of 110 HV, along with an ultimate compressive strength of 185 MPa. The hardness observed in the ECAP-compacted alloy was mainly attributed to the dispersion hardening of Y_2O_3 particles [7].

Another interesting paper in the scope of the Special Issue, submitted by Yang et al. [8] deals with the preparation of layered metal matrix composites and characterisation of the reactions between nickel and germanium after the incorporation of a titanium interlayer on germanium (100) substrate. The results show that, after a microwave annealing (MWA), the nickel germanide layers are formed from 150 °C to 350 °C for 360 s (under nitrogen atmosphere) and the titanium interlayer becomes a titanium cap-layer. These results can be useful for potential precursors for the Secure Digital (S/D) cards contact technology in state-of-the-art Ge-based devices.

The last paper, from Shanmugam et al. [9], faces another important technological aspect: the machining of beta titanium alloys. The original research presents the results of the influence of the additive manufactured (AM) tool in electrochemical micromachining (ECMM) on the machining of a beta titanium alloy. It was shown that the additive manufactured tool can produce a better circularity and overcut than a bare tool. The reason for this result has been discussed and was found to mainly be related to the AM-tool's higher corrosion resistance and higher electrical conductivity.

3. Conclusions and Outlook

This Special Issue covered a variety of topics, presenting recent developments in the synthesis, characterisation, joining and processing of lightweight alloys of the three Al, Mg and Ti families. Moreover, some of the results can provide a basis for effective electronic and industrial applications; however, there are still many challenges to overcome.

As a Guest Editor, I really hope that all the scientific results in this Special Issue can contribute to the advancement of future research on lightweight metal and find application in real industrial environments and markets.

Finally, I would like to thank all the reviewers for their valuable contributions and efforts in improving the academic quality of the research published in this Special Issue. I would also like to give special thanks to all staff at the *Metals* Editorial Office, especially to Toliver Guo, Assistant Editor, always prompt, for supporting and facilitating the publication process in this exceptional pandemic situation.

Funding: This research received no external funding.

Conflicts of Interest: The author declares no conflict of interest.

References

1. Sanctis, M.D.; Fava, A.; Lovicu, G.; Montanari, R.; Richetta, M.; Testani, C.; Varone, A. Mechanical Characterization of a Nano-ODS Steel Prepared by Low-Energy Mechanical Alloying. *Metals* **2017**, *7*, 283. [CrossRef]
2. Loboda, P.; Zvorykin, C.; Zvorykin, V.; Vrzhyzhevskyi, E.; Taranova, T.; Kostin, V. Production and Properties of Electron-Beam-Welded Joints on Ti-TiB Titanium Alloys. *Metals* **2020**, *10*, 522. [CrossRef]
3. Ou, K. L.; Chen, C.-C.; Chiu, C. Production of Oxide Dispersion Strengthened Mg-Zn-Y Alloy by Equal Channel Angular Pressing of Mechanically Alloyed Powder. *Metals* **2020**, *10*, 679. [CrossRef]
4. Testani, C.; Barbieri, G.; Di Schino, A. Analysis of Nanoprecipitation Effect on Toughness Behavior in Warm Worked AA7050 Alloy. *Metals* **2020**, *10*, 1693. [CrossRef]
5. Yang, J.; Ping, Y.; Liu, W.; Yu, W.; Xue, Z.; Wei, X.; Wu, A.; Zhang, B. Ti Interlayer Mediated Uniform NiGe Formation under Low-Temperature Microwave Annealing. *Metals* **2021**, *11*, 488. [CrossRef]
6. Nikolić, F.; Štajduhar, I.; Čanađija, M. Casting Microstructure Inspection Using Computer Vision: Dendrite Spacing in Aluminum Alloys. *Metals* **2021**, *11*, 756. [CrossRef]
7. Shanmugam, R.; Ramoni, M.; Thangamani, G.; Thangaraj, M. Influence of Additive Manufactured Stainless Steel Tool Electrode on Machinability of Beta Titanium Alloy. *Metals* **2021**, *11*, 778. [CrossRef]
8. Huang, S.-J.; Diwan Midyeen, S.; Subramani, M.; Chiang, C.-C. Microstructure Evaluation, Quantitative Phase Analysis, Strengthening Mechanism and Influence of Hybrid Reinforcements (β-SiCp, Bi and Sb) on the Collective Mechanical Properties of the AZ91 Magnesium Matrix. *Metals* **2021**, *11*, 898. [CrossRef]
9. Mortello, M.; Pedemonte, M.; Contuzzi, N.; Casalino, G. Experimental Investigation of Material Properties in FSW Dissimilar Aluminum-Steel Lap Joints. *Metals* **2021**, *11*, 1474. [CrossRef]

Article

Production and Properties of Electron-Beam-Welded Joints on Ti-TiB Titanium Alloys

Petro Loboda [1], Constantine Zvorykin [1,*], Volodymyr Zvorykin [1], Eduard Vrzhyzhevskyi [2], Tatjana Taranova [2] and Valery Kostin [2]

[1]   Igor Sikorsky Kyiv Polytechnic Institute, National Technical University of Ukraine, Prosp. Peremohy 37, 03056 Kyiv, Ukraine; decan@iff.kpi.ua (P.L.); vova.zvorykin@gmail.com (V.Z.)
[2]   Paton Electric Welding Institute of the National Academy of Sciences of Ukraine, Kazymyr Malevych St. 11, 03150 Kyiv, Ukraine; ebw-30@ukr.net (E.V.); taranovatg@ukr.net (T.T.); valerykkos@gmail.com (V.K.)
*   Correspondence: constantine.oleg@gmail.com; Tel.: +380-50-344-5292

Received: 10 March 2020; Accepted: 14 April 2020; Published: 18 April 2020

Abstract: In this article, structural features of Ti-TiB and ($\alpha+\beta$) Ti alloys in the initial state, in the weld and in the heat-affected zone of electron beam welds were investigated. The influences of welding parameters, such as influence of the electron beam velocity, preheating of the welded alloys and the subsequent annealing of the welded joint on the its microstructure, and the mechanical strength and ductility of the critical elements of the joint were studied by scanning electron microscopy using microprobe Auger spectral and X-ray diffraction analysis and tensile tests. It has been shown that the conditions for rapid crystallization of the material from the melt of the weld contribute to refining of reinforcing fibers of TiB and its hardening in comparison with the starting material Ti-TiB. Besides that, influences of the preferential orientation of TiB reinforcing microfibers (along and across the welded butt joint) on the mechanical properties of the welded joint were investigated bz tensile testing. Using the methods of fractographic analysis, the effect of the boron-containing phase on the fracture character of Ti-TiB welded joints was established. It was shown that, along with the strengthening effect, TiB fibers cause embrittlement of the material.

Keywords: titanium alloys; titanium boride; microstructure; mechanical properties; welded joint; electron-beam welding; heat treatment

---

1. Introduction

The development of new titanium alloys for welded constructions is appealing. These alloys could be used in manufacturing engineering, allowing not only achieving high operational characteristics for constructions made from these alloys, but also increased economical effectiveness.

Titanium is characterized by low hardness, a high Young's modulus and relatively low ultimate strength ($\sim$480 MPa). This determines its limited usability in manufacturing in its pure form.

One of effective strengthening mechanisms of metal materials is reinforcement with high-module fibers. Composite metal–ceramic titanium alloys can be reinforced with high-melting compound fibers, such as titanium boride and silicide, which improve the strength by 2–3 times [1–7]. Such materials can be produced by using powder metallurgy. In such cases, high-melting compound fibers are mixed with titanium or titanium hydride powder or are synthesized during sintering of initially formed pressed parts from mixture of titanium hydride and high-melting powders. Another possibility to produce such strengthened alloys is introducing the strengthening fibers during crystallization of eutectic alloys. This ensures their uniform arrangement.

Strength of such metal-ceramic composites is determined by size and quantity of reinforcing fibers [8,9]. The size and quantity of titanium boride fibers during crystallization from eutectic alloys of Ti-B system [10] are determined, first of all, by the temperature gradient at the crystallization front, the homogeneity of chemical composition in melt volume and the stability of thermal conditions (see Table 1).

For production of metal-ceramic composites with TiB fibers, electron-beam melting [11,12] or crucibleless zone melting [13] is used. Titanium is used as a matrix, and titanium boride as fibers, during the obtaining of a Ti-TiB based alloy [12,13].

Table 1. Mechanical properties of titanium and Ti-TiB alloy

Material	Yield Strength $\sigma_{0.2}$, MPa	Tensile Strength σ_t, MPa	Elongation δ, %	Reduction of Area Ψ, %	Vickers Hardness HV, MPa	References
Technical titanium	350–500	450–550	25	55	1400–2000	[12]
BT1-0	-	295–440	24–25	42–55	-	GOST 26492-85
Ti-TiB	631	882	13	16	3500–3700	[12]

Alloys of Ti-TiB type are cost-effective as they do not have to be alloyed with high-priced additions; besides, high-melting reinforcing fibers ensure high mechanical properties both at normal conditions, and under increased temperatures.

When using fiber reinforced alloys in constructions, similar joints between parts made from theses alloys, and dissimilar joints between fiber reinforced Ti-alloys and other titanium alloys and mild steels are to be produced.

Generally, titanium alloys can be characterized by good weldability with each other [14–17], except for some of (α+β)-alloys. The latter ones have a lower weldability, compared to α-alloys and β-alloys with stable structures, because they are more sensitive to the changes of the welding parameters. The required properties can be achieved through complex thermal treatment, which effects the base material and welded joint in different ways. Welding of titanium alloys is done mainly by electron-beam [18] or arc welding in argon [19]. Nevertheless, in case of welding Ti-TiB alloys, it is necessary to retain the structure of micro-reinforcing by TiB fibers. The new requirements for the weld formation conditions and the heat-affected zone (HAZ) have to be determined.

The present article presents basic options of welding parameter optimization for dissimilar joints of Ti-TiB$_n$ alloys with other titanium alloys, and with other structural materials. The article shows the path to achieving high mechanical properties of welded constructions produced by electron-beam welding.

2. Materials and Methods

As our investigations focused on the influence of heat input during welding on the properties of Ti-TiB alloys [12], to keep the number of variables as low as possible, only one Ti-TiB alloy was investigated. The alloy was obtained by sintering powders of Ti (grade ПТК-1 ТУ14-22-57-92 (PTK-1 TU 14-22-57-92) fractional size 45–100 μm >85%, chemical composition, wt.%: N-0.07%, C-0.05%, H-0.35%, Fe-0.35%, Si-0.10%, Ca-0.08%, Cl-0.003%, Ti—the rest) and TiB$_2$ (fractional size ~5 μm, chemical composition (TU 113-07-11.040-89): Ti $\approx$ 70%, B $\approx$ 30%, Fe < 0.05% and C < 0.1%), which was carried out after mixing of Ti-95% and TiB$_2$-5% powders, pressing at $P = 0.65$ GPa and annealing in the β-area temperature range (sintering start temperature 1000 °C, heating at a rate of 0.03 K·s^{-1} to 1200 °C, 3 h, 10 Pa). Briquettes with a diameter of 30 mm and a height of 20 mm, which were obtained after sintering, were filled into the melting chamber of the electron-beam unit UE-208 (УЭ-208, Pilot Paton Plant, Ukraine). The briquettes were melted under a vacuum of $(7–9) \times 10^{-2}$ Pa in a crucible; then the melt was poured into a water-cooled copper crystallizer with a diameter of 110 mm. In the mold, a directed heat sink was achieved by heating the melt surface with electron beams of two electron guns with a heating spot diameter of 35–40 μm (scanning speed 5 mm·s^{-1}). The surface of the ingot was kept in a molten state, forming a temperature gradient that provided directional crystallization with the formation of boride fibers, as described in [12]. After machining the ingot with removing a 2.5 mm layer, multiple deformation processing was performed on a Skoda 500/350 rolling mill with $\varepsilon = 20\%$ degree of plastic deformation. The final blank thickness for the experimental samples was 10 mm.

The blanks for welding investigations were then extracted by water-jet cutting. The ends of the cut of the samples were ground to ensure parallel joint of the welded surfaces, so that the resulting surface roughness Ra was less or equal to 3.2 μm.

The resulting microstructure of a Ti-TiB alloy, consisting of Ti-matrix with boride fibers, can be seen in Figure 1. The quantitative ratio of the phases Ti-95%, TiB_2-5% in the experimental samples was controlled by quantitative X-ray phase analysis by the RIR method.

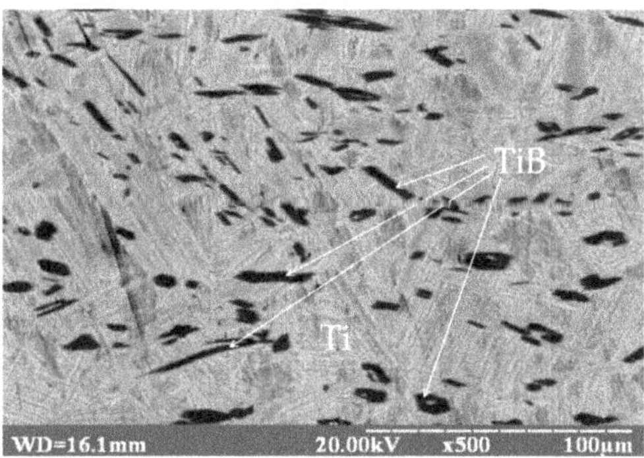

Figure 1. Structure of Ti-TiB melt before welding.

The metallographic analysis showed that TiB microfibers were uniformly distributed over the entire volume of the titanium matrix (α-Ti); their thickness was 2–7 μm (see Figure 1); and the fiber length ranged from 8 μm up to 70 μm. The thickness of TiB fibers was correlated with their average length in a ratio of 1:3.

Ti-TiB can be classified as an alloy, as all of its components go through the stages of melting and subsequent crystallization with the formation of a structural reinforcing TiB phase in the form of reinforcing fibers.

In order to obtain the welded samples on bases of Ti-TiB alloy, electron-beam welding was carried out. Samples in the form of a parallelepiped with size 50 mm × 100 mm × 10 mm were fixed in a clamp and welded at a longer face. The welding was carried out in the following mode of operation: U_{acc} = 60 kV, I_{eb} = 90 mA. Electron beam movement velocity was changed as v_{eb} = 7, 10 and 13 mm·s^{-1}. For welding titanium alloys, the ellipse of the electron beam 3 mm × 4 mm was located with its shorter diameter transverse to the welded joint. The scanning frequency of the electron beam was 170 Hz. The focus was on the weld surface. The diameter of the focused beam was 0.8 mm. The focusing lens current was 960 mA. A lanthanum-boride (single-crystal LaB_6) T-shaped cathode with a diameter of 3 mm was used. The distance from the electron gun to the weld butt was 70 mm. Electron-beam welding was carried out on UL-144 (УЛ-144, Pilot Paton Plant, Ukraine) welding machine.

The thermal source, ensuring the melting of contact zone of welded alloys, was the electron beam. Said beam was formed by an electron gun and was focused onto joint area of welding materials. Under close density of welding materials, the electron beam penetrates into the subsurface area in which they dissipate their energy. The depth of such a layer depends on the accelerating voltage and density of material to be processed, and may be evaluated by formula:

$$\Lambda = 2.35 \times 10^{-12} U_{acc}^2 / \rho, \tag{1}$$

where Λ is electrons penetration depth in μm, and ρ is density of material subjected to welding in g·cm^{-3}. Thus, for a titanium alloy with density of 4.5 g·cm^{-3} under U_{acc} = 60 kV, the value of $\Lambda \approx 20.5$ μm–calculated using the Equation (1). Effective efficiency factor during the electron-beam welding has the values ≈0.85–0.95.

In investigations presented in this article, the power density was ~10^6 W·cm^{-2}. In this power density range the electron-beam energy impact is characterized by melting penetration; the penetration depth to width ratio was 10:1 and more. The melting of the welded materials was complete with the formation of a vapor-gas channel at the point of interaction of the electron beam with the metal surface. The maximum metal temperature in the melt zone was ≈2200–2300 °C [14], which is much higher than melting temperature of titanium and TiB [20].

Specimens of the welded joints' cross sections of welded joins for metallographic investigations were extracted from the welded joints, using a precision water jet cutter KGA 2-R-2500 (Private Enterprise "Roden", Ukraine) by cutting across the weld, with subsequent grinding and polishing. Specimens were then etched with a solution 15%HF + 55%H_2O + 30%HNO_3. Prepared surface layers were photographed and analyzed using a scanning electron microscope (by the JSM-840, JEOL Ltd., Japan) and REM-106I (РЭМ-106И, OJSC "SELMI", Ukraine, Sumy) scanning electron microscopes) and a probe for micro X-ray spectral and Auger spectral analysis with up to ×5000 magnification. The pictures were obtained from secondary (SE) and back-scattered (BSE) electrons. The JAMP-9500F (JEOL Ltd., Japan) scanning electron microscope with the OXFORD EDS INCA Energy 350 (OXFORD INSTRUMENTS INDUSTRIAL PRODUCTS LIMITED, Abingdon, Oxfordshire, UK) energy-dispersive Auger-spectrometer was also used.

Tensile tests were carried out on a ZD-4 (VEB WERKSTOFFPRUFMASCHINEN, Germany, Leipzig) tensile testing machine according to the standard GOST 1497-84. Specimens for tensile test were water cut from the welded plate perpendicular to the weld nut axis so that the weld nut was in the middle of the tensile specimen and then machined to achieve the geometry shown in Figure 2. At least 6 samples were used for confirmation of results repeatability in the experimental cycle.

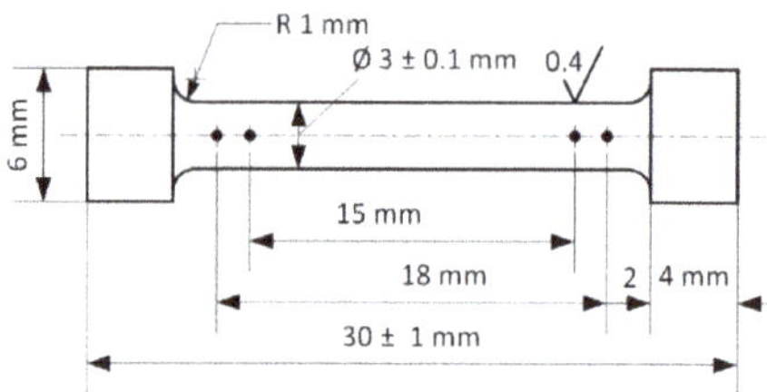

Figure 2. Tensile test specimen sketch-test specimens were manufactured according to GOST 1497-84 (ISO 6892-84).

X-ray structural investigations of the samples were carried out by the DRON-UM-1 (ДРОН-УМ-1, IC "Bourevestnik", Russia) X-ray diffractometer in copper K_α-radiation by step-scan method [12].

3. Results and Discussion

3.1. Welding of Similar Joints Ti-TiB Alloy

3.1.1. Influences of Welding Parameters on Microstructure and Mechanical Properties

In Ti-TiB alloy, the TiB phase, which is known to embrittle the material, leads to a hardening and forms the composite microstructure. This explains the expediency of retaining of microstructure with TiB reinforcing fibers in the welding zone being formed. In the investigations [12,13], it was shown that under conditions of zone melting the reinforcing TiB fibers are formed predominantly elongated in direction of crystallization-front movement. In overheated areas diboride inclusions are formed, which are more faceted and which capture the boron necessary for formation of titanium monoboride reinforcing fibers. This is essential during the selection of parameters of heat input into the weld.

Parameters of electron-beam welding were variated, as shown in Table 2; their influences on microstructure and mechanical properties were determined.

Table 2. Characteristics of the materials of Ti-TiB welded joints produced under various modes of electron-beam welding.

No. of Sample Series	Electron Beam Movement Velocity, v_{eb}, mm·s^{-1}	Initial Temperature of Welded Materials, T_3, °C	Logged Phases	Crystal Lattice Distortion, Δ, %	Average Size of Crystalline Blocks, Π, 10^{-10} m	Mechanical Characteristics				Material Metallographic Structure in Welded Seam Area
						Yield Strength $\sigma_{0.2}$ MPa	Tensile Strength σ_t Mpa	Elongation δ %	Reduction of Area Ψ %	
1	7	20		α Ti-0.1 TiB-0.209	α Ti-150 TiB-56.36	957.2	1040	2.0	6.5	
2	10	20	α Ti TiB	α Ti-0 TiB-0.23	α Ti-142 TiB-195.1	-	1023.1	-	-	
3	13	400		α Ti-0 TiB-0.004	α Ti-141 TiB-84.03	-	950.2	-	-	
4	13	20		α Ti-0.2 TiB-0.41	α Ti-140 TiB-86.65	-	950	-	-	

For the material of welded joint, three typical zones can be defined (see Figure 3):

- Base material hardly influenced by welding process;
- Weld metal which is formed from the melt;
- Transient zone which is formed during the melt's interaction with the base material.

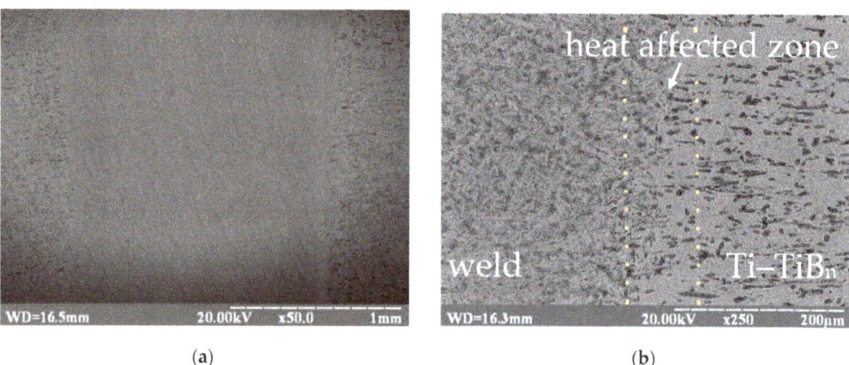

(**a**)   (**b**)

Figure 3. Welded joint produced by electron-beam welding of Ti-TiB$_n$ samples: (**a**) overview; (**b**) microstructure of weld metal, heat-affected zone (HAZ) and base metal.

The metallographic analysis shows that the structure of base material and that of the one obtained in result of welded seam formation differ considerably (see Figures 1 and 4).

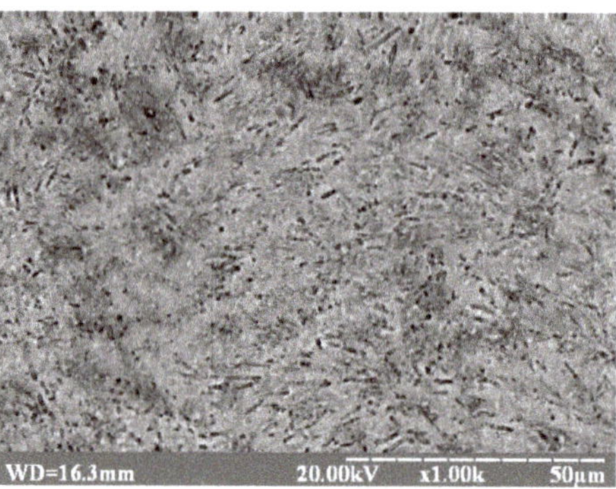

Figure 4. Typical structure of Ti-TiB alloy in the weld metal after welding.

In the welded seam zone, significant changes in structure of welded material were observed (Figure 4), compared to the unaffected base metal (see Figure 1). The microstructure of welded seam material has patterns typical for formation under conditions of fast crystallization of Ti-TiB melts (Figure 5).

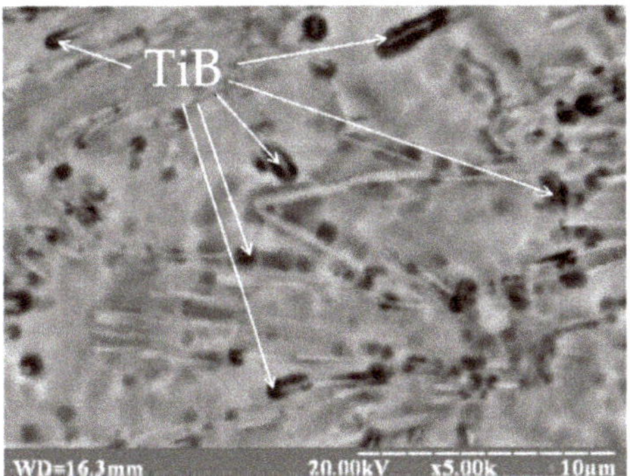

Figure 5. Distribution of boron-containing fibers in weld metal.

The main characteristics of the material revealed in the welded seam area, can conditionally divide the area into three zones (see Table 3).

Table 3. Characteristics of Ti-TiB$_n$ structure (by results of phase X-ray structural investigations, $n = 1$) in the welded seam zone.

No. of Sample Series	Heat Affected Zone (See Figure 6). Area, Retaining the Structural Peculiarities Both Base Material, and Weld Metal	Weld Metal (See Figure 7)	Base Material
1	Zone width is 50 μm on average. In the material are areas without fibrous inclusions, and areas engaging both initial large TiB fibers, and thin micron and submicron fibers.	TiB fibers have not predominant orientation (see Figure 6a). Fibers length is from 2 to 8 μm. TiB fibers thickness is correlated with their length in proportion from 1:4 to 1:8 on average. Average distance between fibers in Ti matrix is 1.5 μm.	It is typical the presence of large elongated grains of titanium boride, for which is typical the decrease of peak intensity of titanium characteristic X-ray emission by 1.2–1.3 times in comparison with the main titanium matrix. Fibers are distributed uniformly by whole volume (their thickness is 2–7 μm (see Figure 2)), fibers with length from 8 to 70 μm are observed. TiB fibers thickness is correlated with their length in proportion 1:3 on average, at that, this proportion for various inclusions is vary from 2:3 to 1:15. Orientation of reinforcing fibers is initial.
2	Zone width is 40 μm on average. In the material are areas without fibrous inclusions, and areas engaging both initial large TiB fibers, and thin micron and submicron fibers.	TiB fibers have not predominant orientation (see Figure 6b). Fibers length is from 1 to 4 μm. TiB fibers thickness is correlated with their length in proportion 1:5 on average, at that, this proportion for various inclusions is vary from 1:4 to 1:6. Average distance between fibers in Ti matrix is 0.8 μm.	
3	Zone width is 32 μm on average. Feature of structure in such transient zone in comparison with samples Nos. 1, 2 and 4 is more high degree of homogeneity which is approaching to the characteristics of welded seam material.	TiB fibers have not predominant orientation (see Figure 6c). Fibers length is from 5 to 20 μm. TiB fibers thickness is correlated with their length in proportion 1:9 on average, at that, this proportion for various inclusions is vary from 1:4 to 1:11. Average distance between fibers in Ti matrix is 4 μm.	
4	Zone width is 28 μm on average. In the material are areas without fibrous inclusions, and areas engaging both initial large TiB fibers, and thin micron and submicron fibers.	TiB fibers have not predominant orientation (see Figure 6d). Fibers length is from 2 to 6 μm. TiB fibers thickness is correlated with their length in proportion 1:10 on average, at that, this proportion for various inclusions is vary from 1:6 to 1:10. Average distance between fibers in Ti matrix is 0.8 μm.	

The results obtained testify the increasing of zone width, formed during interaction of the melt with the sample crystal metal, both under decrease of electron beam movement velocity, and under increase of temperature of the welded samples. This is evidence of the development of relaxation thermodynamic processes in the boundary zone. Completeness of their passing demands availability of a long-time interval and increased temperatures.

The initial temperature and movement velocity of the electron heating source affect the development of relaxation processes in a different way. Under a preheating temperature of 400 °C, a considerable increase of boron-containing phase homogeneity distribution in transient zone and an increase of fiber length and thickness were observed, compared to non-preheated samples. Decrease of the electron beam movement velocity led to a decrease of fiber dimensions, and their distribution in the transient zone became more heterogeneous and had dendritic character.

The metallographic analysis showed that the structures of the initial material and the one obtained in the result of welded seam formation differ considerably (see Figures 6 and 7).

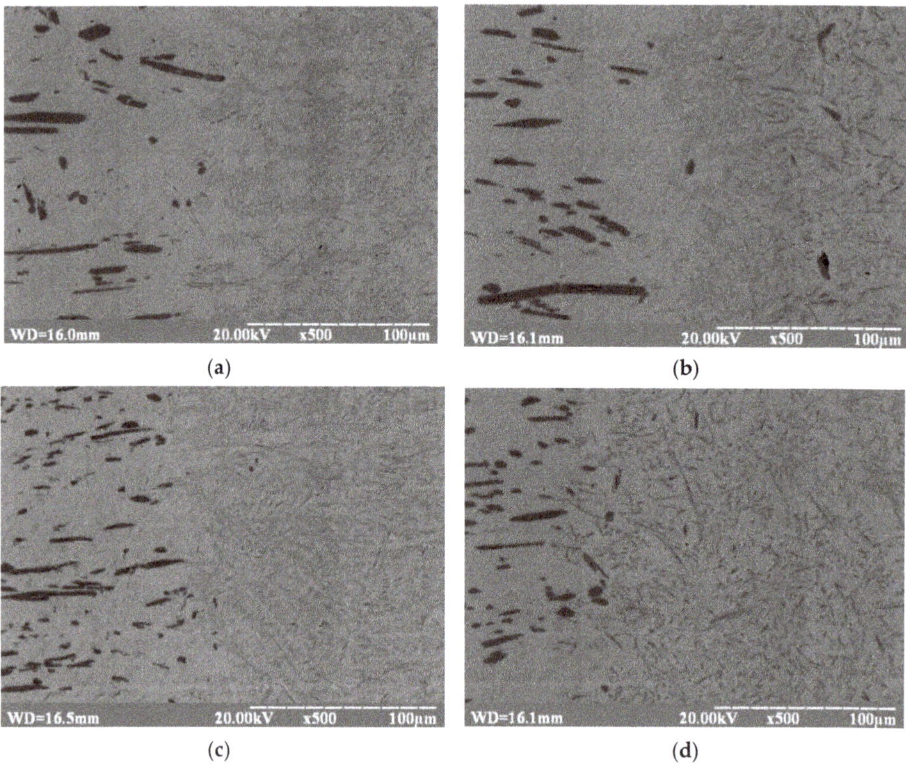

Figure 6. Structure of the material in the area of base material transition into weld metal: (**a**) sample of number 1 series; (**b**) sample of number 2 series; (**c**) sample of number 3 series; (**d**) sample of number 4 series.

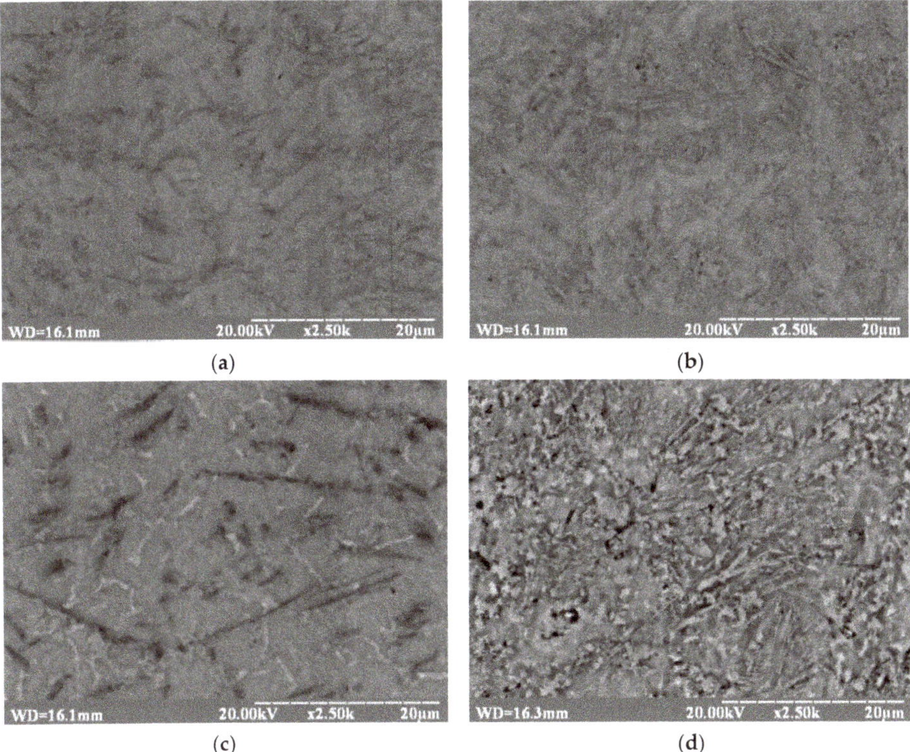

Figure 7. Structure of Ti-TiB alloy in the weld metal after electron-beam welding: (**a**) sample of number 1 series; (**b**) sample of number 2 series; (**c**) sample of number 3 series; (**d**) sample of number 4 series.

In the weld metal, formed by crystallization from the melt, similar trends are observed (see the Figure 7). At the same time, the smallest dimensions and largest uniformity of TiB fibers distribution in TiB matrix are observed in weld metal at medium value of electron beam movement velocity v_{eb} = 10 mm·s^{-1}. Preheating of parts subjected to welding results in considerable increase of TiB fibers' dimensions, which is due to increase of total specific internal energy of material in the zone. Obviously, the increase of v_{eb} results in a considerable increase of temperature gradients, and correspondingly, in essentially greater non-equilibrium of phase and structural states. During v_{eb} increase, the following changes of material structure in the transient zone were observed:

- Segregation of boron-containing fibers in areas with their increased content and formation of areas depleted with such phases (see Figure 6a).
- Growth of boron-containing fibers both by length, and by thickness, upon increasing of residual temperature of joint subjected to welding both at the expense of the electron beam impact increase (see Figure 7a), and at the expense of preliminary heating (see Figure 7c). At that, the effect of preliminary heating is displayed to a considerably greater extent.

Results of mechanical breaking tests of Ti-TiB welded samples demonstrated that material structure in weld metal, and mechanical properties of welded joints and base material, depend on welding parameters (see the Table 2). Particularly, at the minimum velocity of electron beam movement, it was possible to obtain some level of plasticity.

3.1.2. Fractographic Investigations

Fractographic investigations and analysis of fracture zone location regarding to the welded seam of Ti-TiB series welded joints have demonstrated that all samples were fractured by base material. This is evidence of the fact that the strength of welded joint is not less than the strength of the base material (see Figure 8).

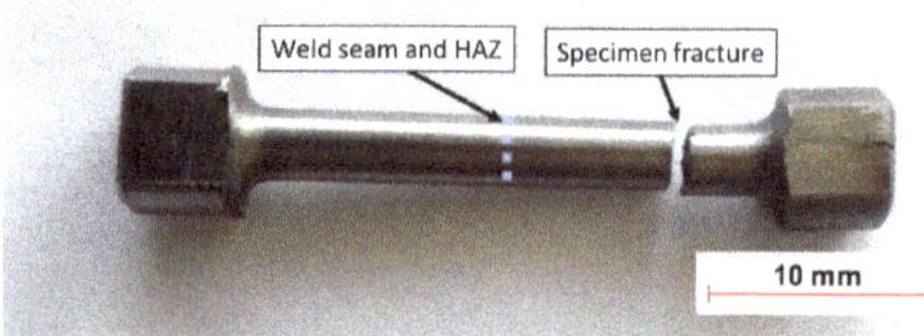

Figure 8. General view of fractured Ti-TiB welded specimen.

During tensile tests of Ti-TiB welded joints it was revealed that under a welding temperature change from 20 to 400 °C the maximum strength and plasticity were observed with the minimum velocity of electron beam movement. At that, the mixed character of the fracture was observed. Fragments of ductile fractured, light-colored wave combs (see Figure 9) were observed on the facture surface, and the fracture was mainly of trans-crystalline type, as determined by its brittle cleavage mechanism.

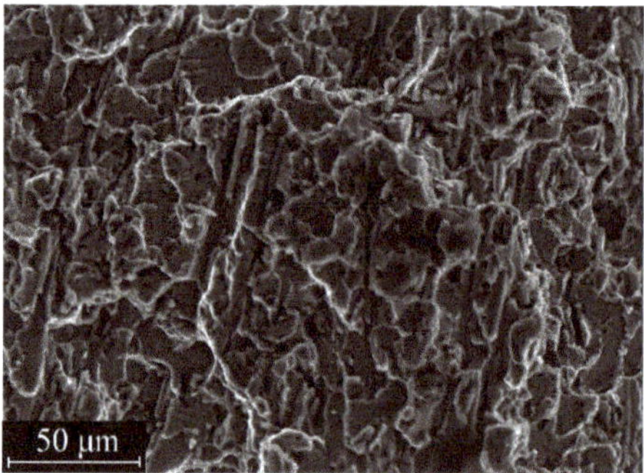

Figure 9. Fractographic investigation of Ti-TiB titanium alloy welded joint produced by electron-beam welding, ×500.

In Figure 9 the fracture area is presented, in which one can see that large rod borides are cracked because of high stresses arising around them, which results in brittle fracturing. Shorter borides (less than 5 μm) do not result in strong localization of stresses, and thus, do not significantly affect the fracture toughness of a plastically deformed matrix. In the Figure 9 the fragment with brittle fracturing is presented, wherein the secondary cracks are revealed on the fracture's surface, which are localized frequently in area near boride inclusions.

Because the welded construction fracturing takes place via the base material, the influences of varied welding parameters may be attributed to the thermal influence on the base material. Minimum distortion of the titanium boride crystal lattice and minimum dimensions of TiB crystalline blocks are observed at minimum velocity of electron beam movement (see the Table 2). Lesser distortion of titanium boride crystal lattice is observed under preliminary heating only, but it results in a considerable increase of mean TiB crystalline block size. According to the results of [10], high adhesive strength of interphase boundaries between titanium matrix and TiB filamentary crystals in VT18U/TiB composites is retained up to the temperatures T = 600–700 °C. Initial heating up to 400° C may result in exceeding these temperatures in the heat-affected zone, and as a consequence, the breaking of adhesion bonds in some areas of large TiB crystalline blocks along with occurrences of additional defects, which promote brittle fracture and strength lowering.

3.2. Welding of Dissimilar Joints between Ti-TiB and (α+β)Ti Alloy

Influences of Welding Parameters and Heat Treatment on Microstructure and Mechanical Properties

In case of Ti-TiB welding with titanium alloy without reinforcing fibers, it is necessary to understand the way in which the transient zone of welded seam shall be formed and will it lead to lowering of its strength.

The (α+β)Ti alloy, in terms of its composition (Al-3.5%, Nb-3.0%, Fe-2.5%, V-1.9%, Mo-1.4%, Zr-1.3%, Si-0.1%, T–the rest), is close to the T110 alloy (5.0–6.0% Al, 3.5–4.8% Nb, 1.5–2.5% Fe, 0.8–2.0%, 0.8–1.8% Mo, 0.3–0.8% Zr, 0.09% O_2, 0.02% N_2, 0.003% H_2), which was developed jointly with the Paton Electric Welding Institute of National Academy of Sciences of Ukraine and "Antonov" State Enterprise (patent of Ukraine number 40087C2 of 16.06.2003) and which is characterized by sufficiently high mechanical properties.

For alloys of Ti-TiB type, the anisotropy of mechanical properties, determined by directionality of reinforcing fibers, is typical. In the Table 2 the mechanical properties of welded joints of Ti-TiB with T110 type alloy are presented, in which the initial orientation of TiB fibers in Ti has a predominant directionality that is perpendicular to the surface subjected to welding. It permits us not only to analyze the processes of production of Ti-TiB and the T110-type alloy welding joint, but also to carry out the analysis of the effects of the reinforcing fibers' orientation in the initial Ti-TiB alloy on the properties of the welded joint produced.

The welding was carried out in the following mode of operation: U_{acc} = 60 kV, I_{eb} = 90 mA, electron beam movement velocity v_{eb} = 7 mm·s^{-1}, beam sweep—elliptic, transversal (3 mm × 4 mm). After welding, some of the welded samples were subjected to annealing for 1 h at the temperature 750 °C (in air condition) or at the temperature 850 °C (in vacuum). Results of mechanical tests of Ti-TiB–T110 samples before and after thermal treatment are in the Table 4.

Table 4. Mechanical characteristics of welded joints of Ti-TiB alloy with T110 titanium alloy obtained by electron-beam welding.

| No. of Sample Series | Kind of Welding Joint | Thermal Treatment | Mechanical Characteristics | | | | Structure of Material in Weld Metal | Notes |
			Yield Strength $\sigma_{0.2}$ MPa	Tensile Strength σ_t MPa	Elongation δ %	Reduction of Area Ψ %		
5		Without annealing	918.9	991.5	1.2	2.3	Reinforcing fibers along the load axis	Fracture by T110 alloy
6	(Ti-TiB)–T110	Without annealing	-	931.3	-	-	Reinforcing fibers across the load axis	Fracture by (Ti-TiB) alloy
7		Annealing 750 °C 1 h (air)	928.7	970.5	2.0	5.9	Reinforcing fibers are across	Fracture by (Ti-TiB) alloy
8		Annealing 850 °C 1 h (vacuum)	-	975.7	-	-	Reinforcing fibers are across	Fracture by (Ti-TiB) alloy

Regarding the beginning of the plastic deformation process before the fracturing of number 7 series samples, the results of fractometric analysis of fracture surfaces are in Figure 10. Peculiarities of the fracture surfaces of series 6 samples, characterized by σ_t = 931.3 MPa (see Figure 10a), and series 6 samples, characterized by σ_t = 970.5 MPa and $\delta \approx 2\%$ (see Figure 10c), permits us to confirm that in both cases the brittle fracture was initiated by developments of brittle cracks at the boundary TiB fiber–titanium matrix. This statement is supported by the fact that TiB fibers are found on all fracture surfaces (see Figure 10b,d). An increase of the specific area of adhesion contact of TiB fibers with a titanium matrix within the total cross-sectional area of a specimen with transversal fiber orientation subjected to deformation, results in an increase of brittle-crack-nucleus development probability in such areas. On the fracture surfaces of series 7 samples, some traces of plastic deformation may also be observed (see Figure 11), which are absent in series 6 and 8 samples.

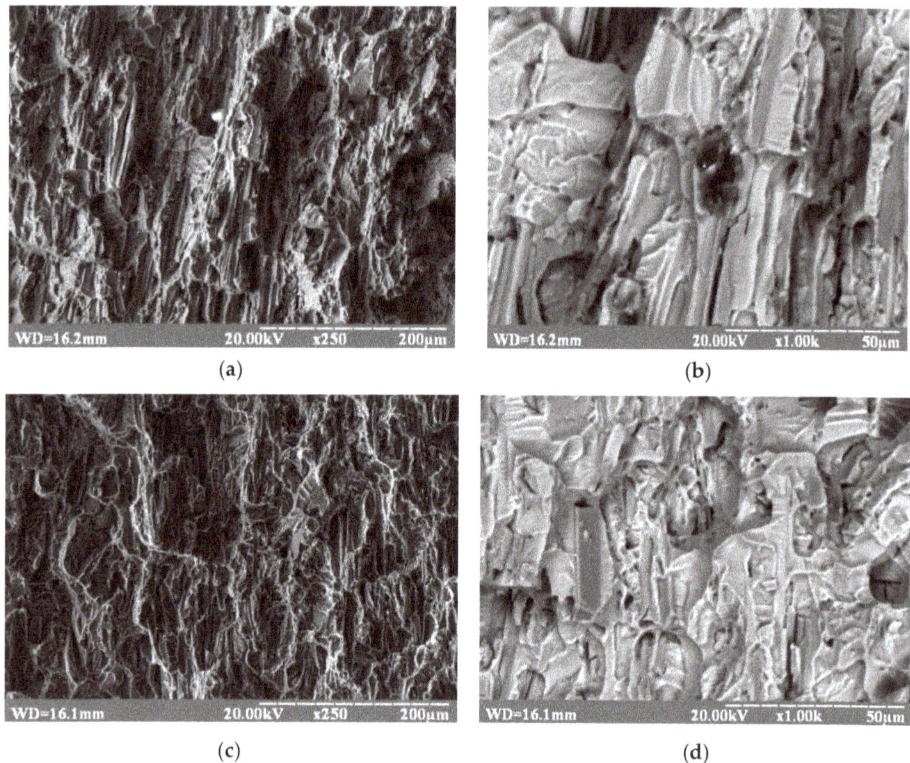

Figure 10. Structure of fracture surfaces of series 6 (**a**,**b**) and 7 (**c**,**d**) samples.

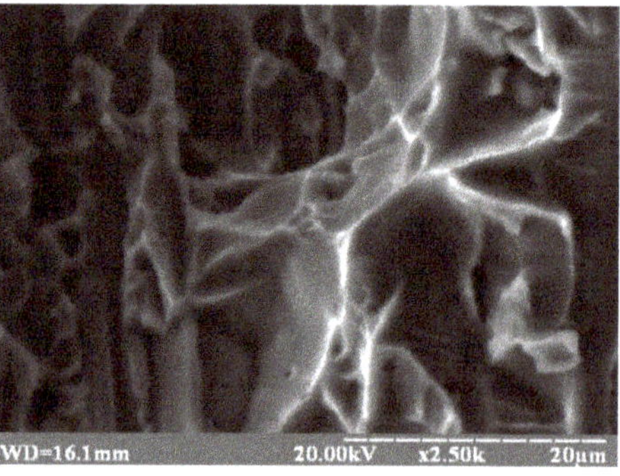

Figure 11. Cellular traces of plastic deformation on fracture surface of a series 7 sample.

In the welded samples, dark colored boron-containing inclusions were observed in the crystallization zone in which molten material contacted Ti-TiB alloy, which retained a solid state (see

Figure 12a). These inclusions have dimensions from submicron quasi-spherical, segregated into the boundary clusters, up to micron, which are close to initial TiB reinforcing fibers.

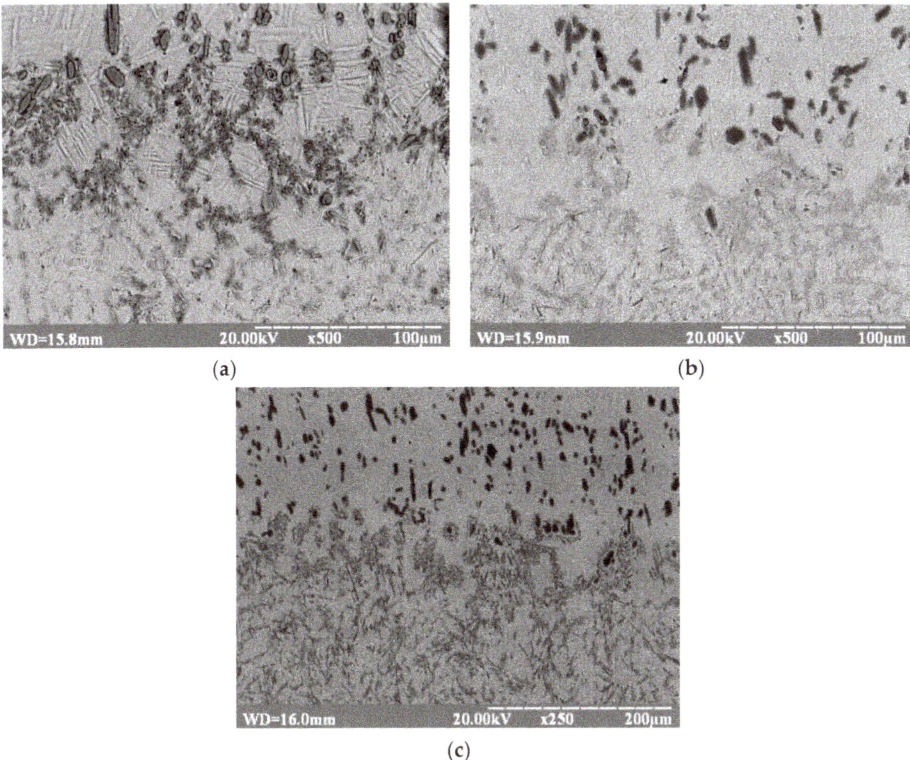

Figure 12. Microstructures of (Ti-TiB)–T110 welded joint material in Ti-TiB—welded seam zone: (**a**) number 6 series sample; (**b**) number 7 series sample; (**c**) number 8 series sample.

Annealing of specimens at 750 °C (see Figure 12b) and 850 °C (see Figure 12c) results in dissolution of the above-mentioned quasi-spherical clusters, while retaining the part of initial reinforcing TiB fibers (see Figure 13b). In such a boundary zone, the submicron TiB-reinforcing fibers are also formed (see Figure 14). After annealing at 750 °C they are distributed more uniformly and after annealing at 850 °C they form a cluster network, typical for all welded seam material (see Figure 15). Dissolution of boron-containing clusters in boundary zone of Ti-TiB-weld metal during subsequent short-term annealing can be considered a proof of thermodynamic instability of these formations. In the experiments [21] the presence of meta-stable Ti_2B was observed. It formed from Ti-B liquid phase at ~2200 °C and subsequently disappeared under conditions of thermodynamic stabilization at ~1800 °C. Very likely, under the conditions of electron-beam welding, rapid cooling at the boundary between Ti-TiB and molten metal resulted in fast crystallization in this zone and the formation of a meta-stable phase, which was dissolved during annealing at 750 °C and 850 °C (1 h).

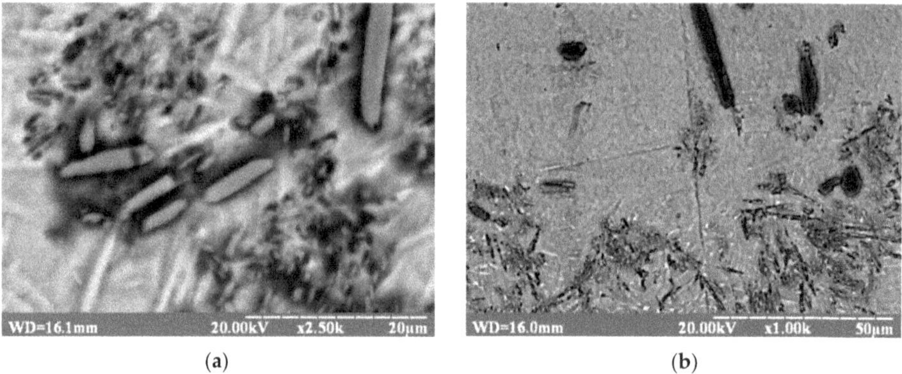

(a)   (b)

Figure 13. Boron-containing inclusions in the Ti-TiB welded seam zone of (Ti-TiB)–T110 welded joint: (**a**) after electron-beam welding (number 6 series sample); (**b**) after electron-beam welding and subsequent annealing at 850 °C (number 8 series sample).

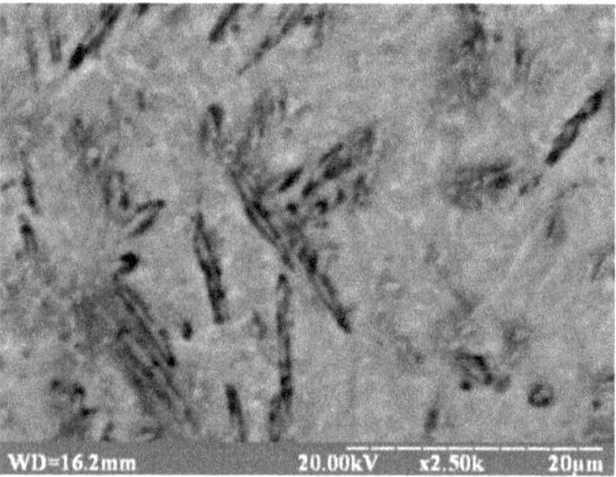

Figure 14. TiB submicron reinforcing fibers of (Ti-TiB)–T110 welded joint area in Ti-TiB—welded seam zone—after annealing at 750 °C.

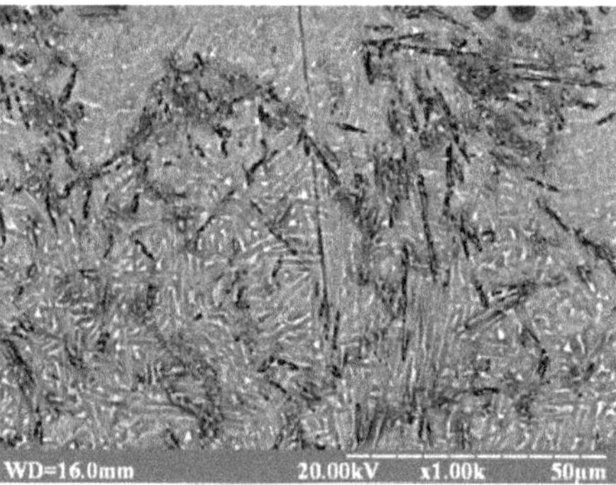

Figure 15. Network of TiB-reinforcing fiber clusters in Ti-TiB—welded seam zone (T110)—after annealing at 850 °C.

With a distance increase from Ti-TiB's welded seam zone towards T110, the effect of annealing on structure of welded seam material was not observed. Furthermore, network cells of submicron boride fiber clusters were enlarged gradually in size (see Figure 16) and were not observed in the zone wherein T110 alloy remained in solid state. In the welded joint the boride submicron fibers do not have predominant directions in the welded seam material, but they are locally oriented predominantly along boundary of their clusters network.

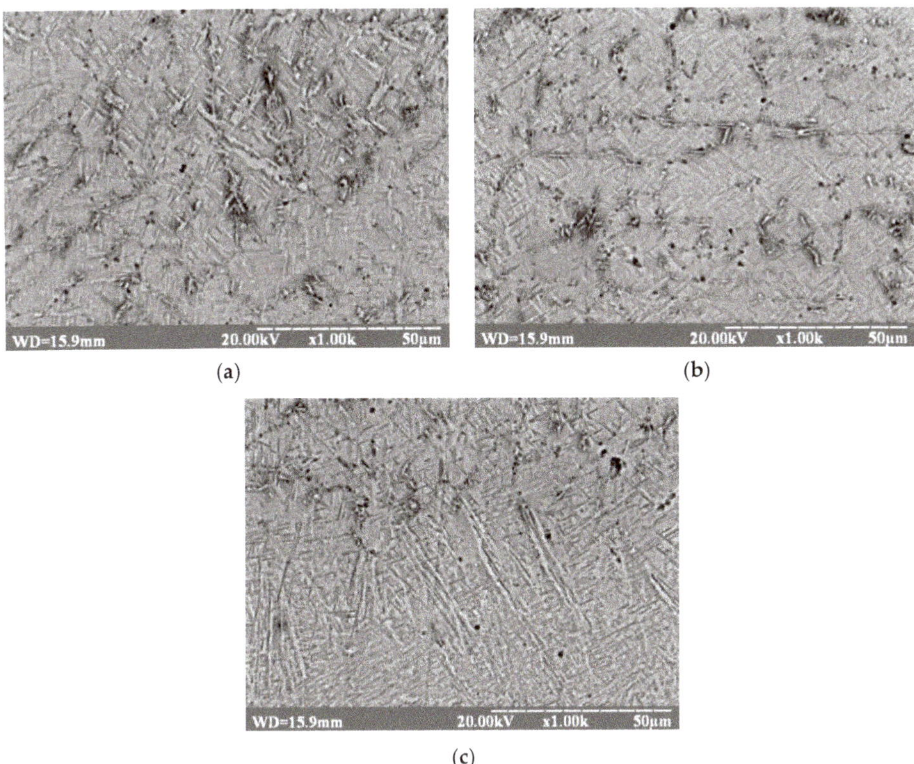

Figure 16. Structure of boride fibers' submicron cluster network with distance from the Ti-TiB—welded seam zone: (**a**) welded seam zone located from the direction of Ti-TiB—welded seam; (**b**) welded seam zone located in the middle area; (**c**) welded seam zone located at the boundary T110 alloy—welded seam (T110 alloy is in the lower part).

X-ray spectral microanalysis of reinforcing fibers in the welded seam zone did not lead to revealing such doped elements in fibers, which are typical for the T110 alloy, besides Al up to 0.18% and Fe up to 0.1%.

During tensile tests conducted at the temperature 20 °C, of the welded Ti-TiB samples (which had an orientation of reinforcing fibers predominantly perpendicular to the plane of the weld butt joint) and T110, fracturing occurred by the base material (Figure 17). With this orientation of the reinforcing fibers, the Ti-TiB alloy was stronger than the T110 alloy, in which fracture occurred. In these cases, the maximum strength with minimum plasticity was observed at the minimum velocity of electron beam movement. The highest level of plasticity was observed at the maximum velocity of electron beam movement and welding of materials initially heated up to 600 °C.

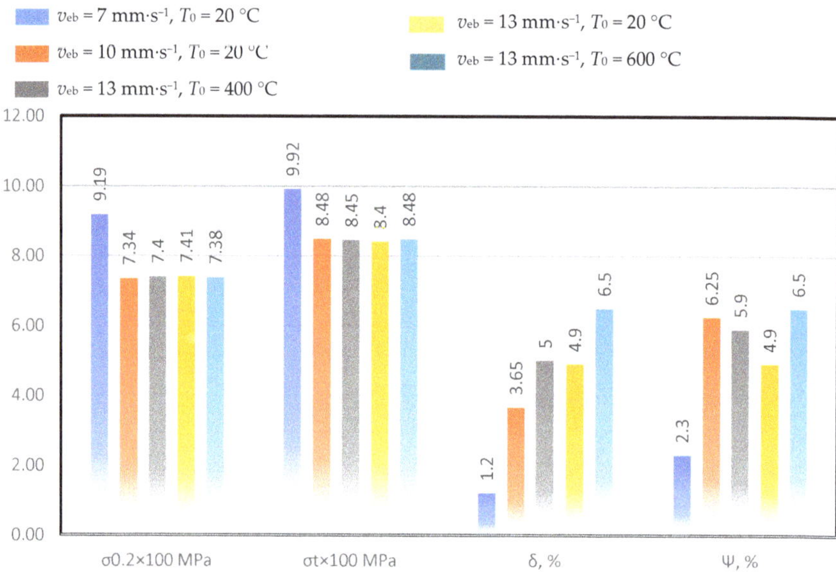

Figure 17. Mechanical properties of (Ti-TiB)–T110 welded joint material obtained under various modes of electron-beam welding.

In order to calculate the strength of the T110 type alloy, by which the fracturing of welded joints took place, the following formula proposed in [22] was used:

$$\sigma_t = 235 + 60 \cdot \iota + 50 \cdot \kappa, \tag{2}$$

where $\iota = \%Al + 0.5\%Sn + 0.33\%Zr + 3.8$, $\kappa = \%Mo + 0.56\%V + 1.25\%Cr + 1.43\%Fe + 0.3\%Nb$.

Calculated σ_t for alloy of T110 type is 1051 MPa according to Equation (2). It matched to the data regarding the strength of T110 alloy quite well (σ_t = 1107 MPa [23]), and it permits us to consider that the reduced level of strength of this welded joint element is associated with the necessity of after-welding heat treatment.

During the production of the welded joint between Ti-TiB and T110 alloys in the case of minimum electron beam velocity $v_{eb} = 7$ mm·s^{-1}, the dendritic structure typical for Ti-TiB welded seam is retained (see Figure 18a). Under increasing of electron beam velocity, the boride inclusions form a cellular structure. An increase of cell dimensions was also observed for preheated parts (600 °C) when compared to parts, welded at the starting temperature of 20 °C (see Figure 18).

Enlargement of cells of dendritic-like microstructure upon increase of initial temperature of materials subjected to welding from 20 °C to 600 °C is probably connected with the incomplete thermodynamic stability of boron-containing areas, adjacent to TiB microfibers.

In the case of mechanical testing of welded joints between Ti-TiB and T110 alloys, fracturing took place in T110. This proves the higher mechanical properties of weld seam. On the other hand, it demonstrates the necessity of heat treatment optimization in order to increase mechanical properties of welded joints.

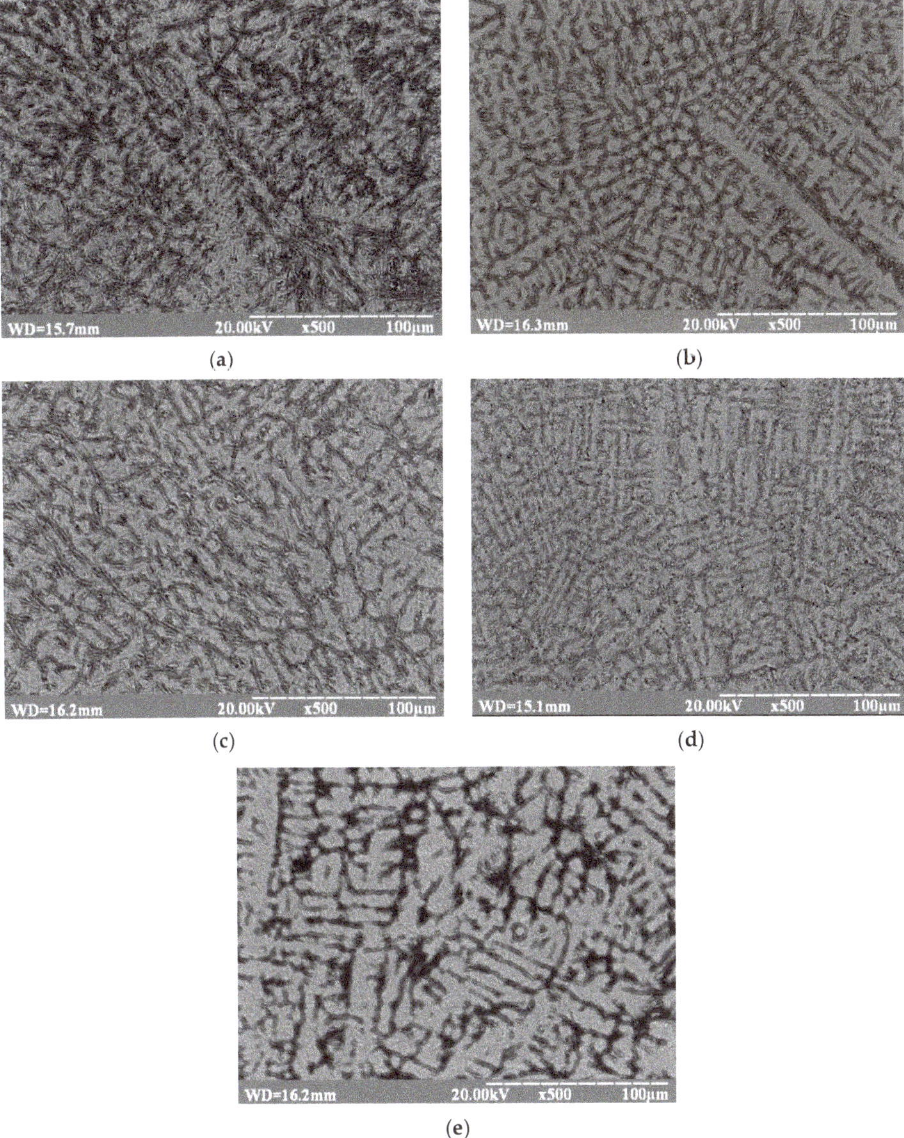

Figure 18. Microstructures of the welded seam material of the (Ti-TiB)–T110 joint: (**a**) $v_{eb} = 7$ mm·s^{-1}, $T_0 = 20$ °C; (**b**) $v_{eb} = 10$ mm·s^{-1}, $T_0 = 20$ °C; (**c**) $v_{eb} = 13$ mm·s^{-1}, $T_0 = 400$ °C; (**d**) $v_{eb} = 13$ mm·s^{-1}, $T_0 = 20$ °C; (**e**) $v_{eb} = 13$ mm·s^{-1}, $T_0 = 600$ °C.

Investigation of surface fracturing of a given series of dissimilar welded joints is shown Figure 19a,b. Elements of ductile fracture (light-colored wave combs) are observed on fracture surfaces, and brittle areas correspond, obviously, to the cellular structure areas.

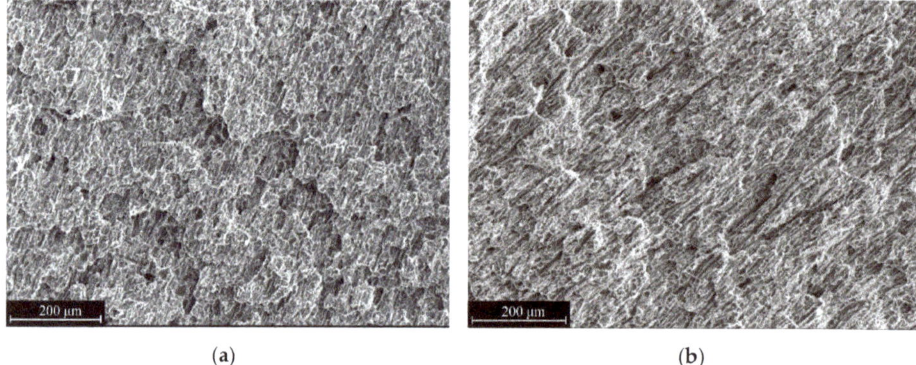

(a)　　　　　　　　　　　　　　　　　　(b)

Figure 19. General views of the fracture surfaces of dissimilar welded joints of (Ti-TiB)–T110 series samples at the temperature 20 °C, ×100: (**a**) plasticity 1.2%; (**b**) plasticity 4.9%.

Fragmentation of borides absent tearing from the matrix (see the Figure 20) points out the high adhesive strength of the boundary between the titanium matrix and TiB particles. The presence of a boride phase with extremely low plasticity may initiate the development and opening of a brittle crack in the fracture process, which is able to propagate across all of the Ti-TiB material.

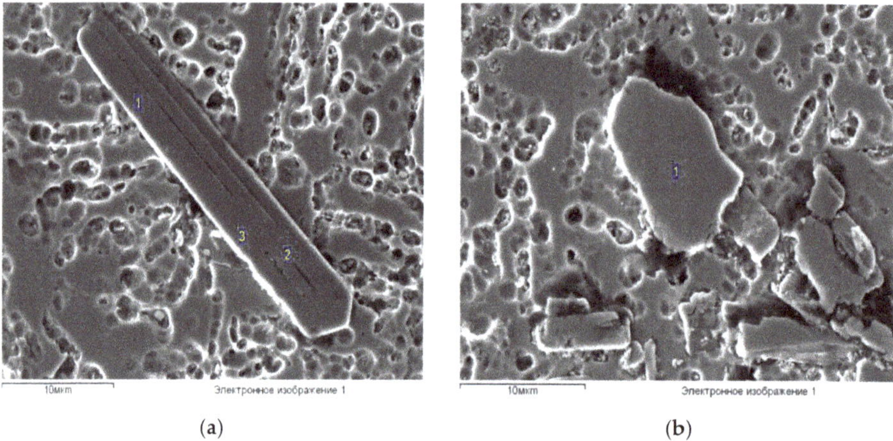

(a)　　　　　　　　　　　　　　　　　　(b)

Figure 20. TiB particles in the deformed titanium matrix of the Ti-TiB alloy of a (Ti-TiB)–T110 welded joint: (**a**) boride plate retained integrity while deforming (TiB$_n$: n = 1.47 (1); 1.01 (2); 1.16 (3)); (**b**) defragmented boride plate (n = 1.06).

During the tensile testing of dissimilar welded joints of Ti-TiB and T110 alloys, produced under low electron beam velocities, a certain level of plasticity was observed: at the minimum electron beam movement velocity of 7 mm·s^{-1} it was 1.2%, and at the electron beam movement velocity of 13 mm·s^{-1} it was 4.9%, but plasticity increase resulted in strength decrease (see the Figure 19).

Figure 20a,b present the general view of fracture surface of welded joints between (Ti-TiB)–T110 series samples for which a decent level of plasticity is observed under all welding parameters. In this case the fracture surface shows elements of ductile fracture (light-colored wave combs) (see Figure 21), which points to the main crack propagation character (see the Figure 19a). As was mentioned above, in all cases of mechanical testing of (Ti-TiB)–T110 series samples the fracture took place in alloy T110 away from weld seam area.

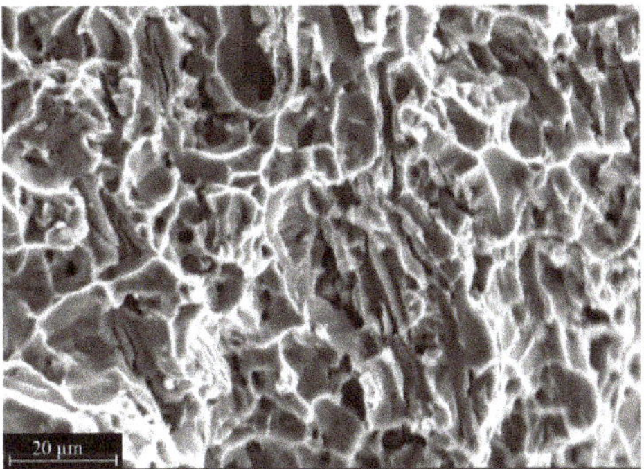

Figure 21. Fragment of ductile fracture of fracture surface of (Ti-TiB)–T110 series samples welded joint at the temperature of subjected to welding samples 600 °C, ×1000.

It should be noted that the conducted fractographic investigations have shown that the fracture surfaces of (Ti-TiB)–T110 series samples demonstrate mixed brittle-ductile fracturing with a prevailing brittle fracture mechanism (see the Figure 22). Increasing of the beam velocity and preheating temperature when welding dissimilar joints, leads to an increase of the ductile fracture fraction. Upon reaching the critical level of stresses in Ti-TiB alloy, large borides are cracked, supposedly due to high stress concentration around them, which results in brittle fracturing (see the Figure 22b). The presence of a larger amount of short borides (less than 5 μm), which are typical for welded seam material, do not result in strong stress localization and thus do not affect the fracture toughness of the plastically deformed matrix.

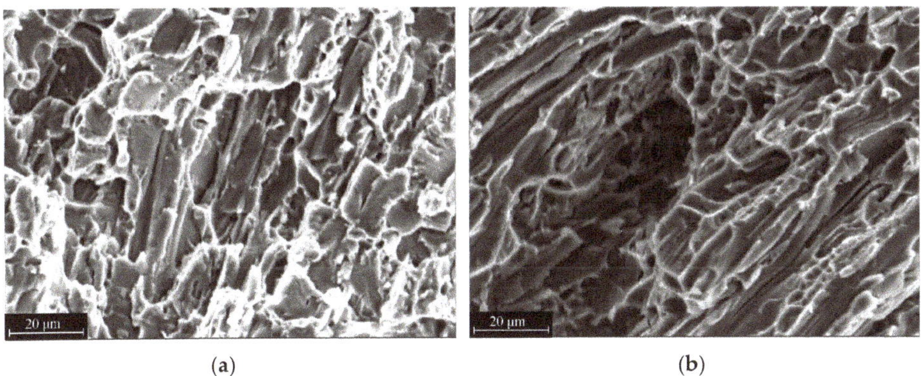

(**a**)   (**b**)

Figure 22. Fragment of ductile fracture of fracture surface of (Ti-TiB)–T110 welded joint, ×1000: (**a**) v_{eb} = 7 mm·s^{-1}; (**b**) v_{eb} = 13 mm·s^{-1}.

As mentioned previously, all the specimens fractured in the base metal. This can be explained by the fact that boride fibers' sizes in the weld metal are significantly smaller (TiB fibers ranging from 1 to 20 μm, see Figures 4, 5 and 13–16, and Tables 2–4) compared to the base metal (TiB fibers are between 8 and 70 μm, as shown in Figure 1). Such refinement of boride fibers could possibly lead to higher strength of Ti-TiB, as discussed by Kaczmarek et al. [7]. However, a systematic study with a focus on fiber size variation and distribution and their influence on the mechanical properties of welded joints with a broader range of TiB fiber sizes is needed to check this assumption.

4. Conclusions

1. The electron-beam welding with $U_{acc} = 60$ kV, $I_{eb} = 90$ mA, beam sweep elliptic (3 mm × 4 mm), transversal, under all values of v_{eb} in the range from 7 to 13 mm·s^{-1} ensures production of welded joints of Ti-TiB alloy samples, containing 5 mas. % of TiB$_2$ in titanium base.

2. During weld seam formation, the basic material is subjected to microstructural changes which result in decreasing of boron-containing fibers thickness from 3–8 μm to 0.1–0.9 μm. Their distinctive initial directionality is lost, and a considerable decrease in the thickness-to-length ratio of the fibers in the boron-containing phase was observed.

3. The transient zone between the base metal and the welded seam material of Ti-TiB alloy has typical dimensions of 40–50 μm. In this zone both the boron-containing fibers of 3–8 μm thickness and 8–40 μm length, typical for an initial material, and thin long boron-containing fibers of 0.1–0.9 μm thickness and 3–15 μm length, typical for a welded seam, were observed.

4. Preheating of welded samples to 400 °C results in a considerable rise of the homogeneity distribution degree of the boron-containing phase in the transient zone and an increase of fiber dimensions of this phase both by length and thickness, both in the transient zone and in the weld metal, compared to non-preheated specimens.

5. Increase of electron beam movement velocity from 7 to 13 mm·s^{-1} results in an increase of the transient zone between base material and welded seam material and in a more uniform distribution of boron-containing fibers in titanium matrix.

6. Upon reaching the critical level of tensile stresses, the welded joints of Ti-TiB and T110 type alloys, produced by electron-beam welding under all values of v_{eb} in range from 7 to 13 mm·s^{-1} regardless of preheating, are fractured outside of the welded seam zone; the fracture surfaces show signs of mainly brittle fractures.

7. Predominant transversal orientation of TiB-reinforcing fibers in Ti-TiB alloy leads to decrease of mechanical properties of dissimilar joints between Ti-TiB and T110. For such joints, the condition of Ti-TiB alloy is critical, as it is subjected to brittle fracturing. The thermal annealing of the welded joint at the temperature 750 °C (1 h) increases the plasticity of this alloy.

Author Contributions: Conceptualization, P.L. and C.Z.; methodology, E.V. and T.T.; software, V.K.; validation, V.K., T.T. and E.V.; formal analysis, V.Z.; investigation, V.Z., V.K. and T.T.; resources, P.L.; data curation, C.Z.; writing—original draft preparation, V.Z.; writing—review and editing, C.Z.; visualization, V.K. and T.T.; supervision, P.L.; project administration, C.Z.; funding acquisition, P.L. All authors have read and agreed to the published version of the manuscript.

Funding: This research received no external funding.

Conflicts of Interest: The authors declare no conflict of interest.

References

1. Gaisin, R.A.; Imayev, V.M.; Imayev, R.M.; Gaisina, É.R. Microstructure and mechanical properties of VT25U/TiB composite prepared in situ by casting and subjected to hot forging. *Lett. Mater.* **2017**, *7*, 186–196. [CrossRef]

2. Attar, H.; Bonisch, M.; Calin, M.; Zhang, L.-C.; Scudino, S.; Eckert, J. Selective laser melting of in situ titanium–titanium boride composites: Processing, microstructure and mechanical properties. *Acta Mater.* **2014**, *76*, 13–22. [CrossRef]

3. Morsi, K.; Patel, V.V. Processing and properties of titanium–titanium boride (TiB) matrix composites—A review. *J. Mater. Sci.* **2007**, *42*, 2037–2047. [CrossRef]

4. Rahoma, H.K.S.; Wang, X.P.; Kong, F.T.; Chen, Y.Y.; Han, J.C.; Derradji, M. Effect of ($\alpha+\beta$) heat treatment on microstructure and mechanical properties of (TiB+TiC)/Ti-B20 matrix composite. *Mater. Des.* **2015**, *87*, 488–494. [CrossRef]

5. García de Cortazar, M.; Agote, I.; Silveira, E.; Egizabal, P.; Coleto, J.; Petitcorps, Y.L. Titanium composite materials for transportation applications. *JOM* **2008**, *60*, 40–46. [CrossRef]

6. Zhang, J.; Ke, W.; Ji, W.; Fan, Z.; Wang, W.; Fu, Z. Microstructure and properties of insitu titanium boride (TiB)/titanium (TI) composites. *Mater. Sci. Eng. A* **2015**, *648*, 158–163. [CrossRef]

7. Kaczmarek, M.; Jurczyk, M.U.; Miklaszewski, A.; Paszel-Jaworska, A.; Romaniuk, A.; Lipińska, N.; Żurawski, J.; Urbaniak, P.; Jurczyk, K. In vitro biocompatibility of titanium after plasma surface alloying with boron. *Mater. Sci. Eng. C* **2016**, *69*, 1240–1247. [CrossRef] [PubMed]

8. Vishniakov, L.R.; Grudina, T.V.; Kadyrov, V.K.; Karpinos, D.I.; Oleinik, V.I. *Composition Materials: Reference Book*; Naukova Dumka: Kiev, Ukraine, 1985; 589p. (In Russian)

9. Gaisin, R.A.; Imayev, V.M.; Imayev, R.M. Microstructure and mechanical properties of Ti–TiB based short-fiber composite materials manufactured by casting and subjected to deformation processing. *Russ. Phys. J.* **2015**, *58*, 848–853. [CrossRef]

10. Elliot, R. *Eutectic Solidification Processing*; Engl. transl.; Shvindlerman, L.S., Ed.; Metallurgy: Moscow, Russia, 1987; 352p. (In Russian)

11. Paton, B.E.; Trigub, N.P.; Akhonin, S.V.; Zhuk, G.V. *Titan Electron-Beam Melting*; Naukova Dumka: Kiev, Ukraine, 2006; 248p, ISBN 966-00-0665-9. (In Russian)

12. Grigorenko, G.M.; Akhonin, S.V.; Loboda, P.I.; Grigorenko, S.G.; Severin, A.Y.; Berezos, V.A.; Bogomol, Y.I. Structure and properties of titanium alloy, alloyed with boron, produced by the method of electron beam remelting. *Electrometall. Today* **2016**, *1*, 21–25. (In Russian) [CrossRef]

13. Loboda, P.I. *Directionally Crystallized Borides*; Praimdruk: Kyiv, Ukraine, 2012; 395p. (In Ukrainian)

14. Gurevich, S.M.; Zamkov, V.N.; Kompan, Y.Y.; Kushnirenko, N.A.; Kharchenko, G.K.; Blaschuk, V.E.; Volkov, V.B.; Zagrebenyuk, S.D.; Prilutsky, V.P.; Sabokar, V.K. *Metallurgy and Technology of Welding of Titan and It Alloys*; Zamkov, V.N., Ed.; Naukova Dumka: Kiev, Ukraine, 1986; 240p. (In Russian)

15. Grigorenko, G.M.; Zadorozhnuk, O.M. Structure, mechanical properties and weldability of pseudo-α and ($\alpha+\beta$)-Ti alloys strengthened by silicides. *Electrometall. Today* **2016**, *2*, 51–56. (In Russian) [CrossRef]

16. Grabin, V.F. *Fundamentals of Metal Science and Thermal Treatment of Welded Joints from Titanium Alloys*; Naukova Dumka: Kiev, Ukraine, 1975; 262p. (In Russian)

17. Shelyagin, V.D.; Khaskin, V.Y.; Akhonin, S.V.; Belous, V.Y.; Petrichenko, I.K.; Siora, A.V.; Palagesha, A.N.; Selin, R.V. Peculiarities of laser-arc welding of titanium alloys. *Autom. Weld.* **2012**, *12*, 36–40. (In Russian)

18. Nazarenko, O.K.; Kajdalov, A.A.; Kovbasenko, S.N.; Bondarev, A.A.; Shevelev, A.A.; Chvertko, A.I.; Zubchenko, J.V.; Lankin, J.N.; Sheljagin, V.D. *Electron-Beam Welding*; Paton, B.E., Ed.; Naukova Dumka: Kiev, Ukraine, 1987; 255p. (In Russian)

19. Akhonin, S.V.; Belous, V.Y.; Antonyuk, S.L.; Petrichenko, I.K.; Selin, R.V. Properties of high-strength titanium alloy T110 joints made by fusion welding. *Autom. Weld.* **2014**, *1*, 54–57. (In Russian)

20. *State Diagrams of Binary Metal Systems*; Ljakishev, N.P. (Ed.) Mashinostroenie: Moscow, Russia, 1996; Volume 1, 992p, ISBN 5-217-02688-X. (In Russian)

21. Palty, A.E.; Margolin, H.; Nielsen, J.P. Titanium-Nitrogen and Titanium-Boron systems. *Trans. ASM* **1954**, *46*, 312.

22. Khorev, A.I. Fundamental and applied works on titanium alloys and prospective lines of their development. *Technol. Mach. Build.* **2014**, *11*, 5–10. (In Russian)

23. Bychkov, A.S.; Moliar, A.G. Operational load-carrying ability of constructional elements of domestic aircraft of transport category from titanium alloys. *Open Inf. Comput. Integr. Technol.* **2016**, *71*, 18–25. (In Russian)

Article

Production of Oxide Dispersion Strengthened Mg-Zn-Y Alloy by Equal Channel Angular Pressing of Mechanically Alloyed Powder

Keng-Liang Ou [1,2,3,4], Chia-Chun Chen [5] and Chun Chiu [5,*]

[1]   Department of Dentistry, Taipei Medical University Hospital, Taipei 110, Taiwan; klouyu@gmail.com
[2]   Department of Dentistry, Cathay General Hospital, Taipei 106, Taiwan
[3]   Department of Dentistry, Taipei Medical University-Shuang Ho Hospital, New Taipei City 235, Taiwan
[4]   3D Global Biotech Inc., New Taipei City 221, Taiwan
[5]   Department of Mechanical Engineering, National Taiwan University of Science and Technology, Taipei 106, Taiwan; chiachun7948@gmail.com
*   Correspondence: cchiu@mail.ntust.edu.tw; Tel.: +886-2-27376479

Received: 8 April 2020; Accepted: 19 May 2020; Published: 21 May 2020

Abstract: Mg-Zn-Y alloys with long-period stacking ordered structures (LPSO) have attracted attention due to their excellent mechanical properties. In addition to the LPSO structure, Mg alloys can also be strengthened by oxide particles. In the present study, oxide dispersion strengthened $Mg_{97}Zn_1Y_2$ (at%) alloys were prepared by equal channel angular pressing (ECAP) of mechanical alloyed (MA) powder under an oxygen gas atmosphere. The 20-h-MA powder had a particle size of 28 μm and a crystallite size of 36 nm. During the MA process followed by ECAP, an Mg matrix with dispersed Y_2O_3 (and MgO) particles was formed. The alloy processed by ECAP exhibited a hardness of 110 HV and a compressive strength of 185 MPa. Compared to pure Mg, the increased hardness was due to the dispersion strengthening of Y_2O_3 and MgO particles and solution strengthening of Zn and Y.

Keywords: powder metallurgy; mechanical alloying; equal channel angular pressing; dispersion strengthening

---

1. Introduction

Research into light-weight materials is essential to improve fuel efficiency and achieve the goal of energy saving. As one of the most important light-weight structural materials in automotive, aerospace and biomaterial industries, magnesium alloys have drawn much attention because of its high specific strength to weight ratio and recyclability [1–4]. However, the conventional AZ (Mg-Al-Zn) and AM (Mg-Al-Mn) series of Mg alloys will not meet the expanded requirements for load-bearing components [5]. As a result, it is necessary to develop high-performance Mg alloys that will satisfy the extended industrial applications. In recent years, studies have shown that WZ (Mg-Zn-Y) series Mg alloy, which has a long period stacking ordered (LPSO) phase, can improve the mechanical properties of the Mg alloys. As for the fabrication of WZ series alloys, it can be performed by applying the powder metallurgy (P/M) route, which combines powder preparation and consolidation of powders. Mg alloy powders can be produced by rapid solidification such as melt spinning and gas atomization while the consolidation of the alloy powders can be performed by severe plastic deformation (SPD) including hot extrusion and high-pressure torsion (HPT) [6]. $Mg_{97}Zn_1Y_2$ (at%) alloys were fabricated by a rapidly solidified powder metallurgy (RS/PM) method [7,8]. The bulk alloy prepared by warm extrusion of gas-atomized $Mg_{97}Zn_1Y_2$ powders exhibited an interesting tensile strength of 610 MPa and modest elongation of 5% in comparison to those of AZ and ZK series Mg alloys prepared by casting. The high strength was attributed to the presence of the LPSO phase as well as the fine Mg matrix with a grain size of 100–200 nm.

Enhancement of the strength of Mg alloys was also achieved by oxide-dispersion strengthening. Lee et al. reported the compaction of pure Mg powders using equal channel angular pressing (ECAP) [9]. It was observed that MgO was formed during the consolidation process. The sample after four passes of ECAP at 300 °C showed the highest ultimate compressive strength of 193 MPa. The enhanced strength was the result of fine and uniformly-distributed MgO particles. Inspired by the formation of oxide during the compaction process, simultaneous compaction of Mg alloys and oxide dispersion processes was proposed in the present study although the source of oxygen might be from an oxygen gas in the reaction chamber rather than from the oxidizer in the starting materials for an internal oxidation process [10–12]. By introducing oxygen during the compaction process and using an alloying element with a higher oxygen affinity than that of Mg, oxide particles could be formed. The key to the successful internal oxidation of Mg-based alloys is the distribution of oxide. If the oxide could be distributed uniformly in the Mg-matrix rather than agglomerated at the grain boundary region after the compaction, the mechanical property could be enhanced by dispersion strengthening [13].

In addition to direct extrusion of P/M powders, the bulk alloy can be produced by equal channel angular pressing (ECAP) of mechanically-alloyed (MA) powders. Mechanical alloying is a solid-state powder processing technique, in which the welding, fracturing and rewelding of powders are repeated in a ball mill machine to synthesize an alloy [14]. Through the MA process, powders containing super-saturated solid solution, non-equilibrium phase, as well as ultrafine or even monocrystalline grain have been obtained. Zhou et al. synthesized nanocrystalline AZ31-Ti composite by cold pressing of MA powder [15]. After milling for 110 h, a nanocrystalline Mg matrix with a crystallite size of 66 nm was formed. The as-milled AZ31 Mg alloy containing 27 wt% Ti had a hardness of 147 HV, which was three times higher than that of the as-cast AZ31 Mg alloy. Fabrication of $Mg_{97}Zn_1Y_2$ alloy powder by mechanical alloying of pure Mg, Zn and Y was reported [16]. It was found that peaks of Y and Zn disappeared after milling for 1 h and peaks belonged to $Mg_{24}Y_5$ were identified after 5 h of milling. The crystallite size of the Mg phase was 27 nm. The LPSO phase or oxide phase was not formed after mechanical alloying. The consolidation of the synthesized powder was not reported.

Equal channel angular pressing, one of the severe plastic deformation (SPD) process, was originally developed to prepare bulk alloys with ultrafine structures. In the initial design, the bulk ingot was used as a starring material. In order to obtain nanometer grains, it has been suggested that powders should be used as starting materials. Thus, ECAP processing for the powder was developed. Bulk alloy or composites have been prepared by consolidating alloy or ceramic reinforced metal matrix composite powders using ECAP [17–21]. It has been demonstrated that bulk Cu material having a grain size of less than 100 nm could be prepared by ECAP [22].

In our previous works, $Mg_{97}Zn_1Y_2$ alloys have been fabricated by stir-casting [23] and by ECAP of mechanically-milled powder (as-cast alloy powder was used as the starting material) under an argon atmosphere [24]. The results showed that the phases existing in the alloy strongly depended on the processing method. $Mg_{97}Zn_1Y_2$ alloy prepared by casting contained α-Mg and $Mg_{12}Zn_1Y_1$ (LPSO phase) while α-Mg and $Mg_{24}Y_5$ were formed in the alloy prepared by combination of ECAP and mechanical milling. Literature reviews revealed that no prior attempts have been made to study oxide dispersion in Mg-Zn-Y alloy during the ECAP process. It would be interesting to fabricate bulk Mg-Zn-Y alloy by ECAP of mechanically alloyed powder under an oxygen atmosphere. Accordingly, a simultaneous compaction of the Mg alloy and oxide dispersion process was proposed. The aim of the present study was to fabricate bulk $Mg_{97}Zn_1Y_2$ alloy by ECAP of as-alloyed powders under an oxygen gas atmosphere and to provide early insight into the properties of the compacted bulk alloy. In the present study, elemental Mg, Zn, and Y powders were used as starting materials for mechanical alloying. Microstructure and mechanical properties of the bulk alloy prepared by ECAP of as-alloyed powders were studied to investigate the effect of the ECAP route and pass number on the compaction of powders.

2. Materials and Methods

Elemental Mg (99.9%, <45 μm), Zn (99.9%, <150 μm) and Y (99.6%, <380 μm) powders were used as starting materials for preparing $Mg_{97}Zn_1Y_2$ (at%) alloy powder by the mechanical alloying (MA) process. The stoichiometric ratio of Mg, Zn and Y powder was 97: 1: 2. The powder mixture was milled for 30 h under argon (99.999%) in a planetary ball mill (RETSCH PM100 RETSCH, Haan, Germany). A rotation speed of 250 rpm and a ball-to-powder ratio of 20 were used. To minimize oxidation of the powder, all of the powder handling was performed inside a glove box (homemade) with an argon atmosphere.

For the equal channel angular pressing (ECAP) process, as-milled powders were filled inside a copper tube and compacted by mechanical force, and sealed with a copper cup under an argon-oxygen gas mixture. The sealed tubes were then processed by ECAP at 300 °C. The deformation routes that can be applied in an ECAP process are summarized in Table 1. In the present study, two types of pressing routes including route Bc and route C were selected to investigate the effects of deformation route on microstructure and mechanical properties of the compacted samples. Details of the experimental parameters for the ECAP process can be found elsewhere [24]. The samples were labeled according to their processing route and number of passes. The samples pressed with one pass, two and four pass using route Bc or C were labeled as 1pass, Bc2, Bc4, C2 and C4, respectively.

Table 1. Four basic routes that can be applied in a ECAP process.

Route	Process
A	The sample is not rotated.
B_A	Rotating the sample around its longitudinal axis by 90° clockwise and counterclockwise alternatively
Bc	Rotating the sample around its longitudinal axis after each pass by 90° clockwise
C	Rotating the sample around its longitudinal axis after each pass by 180° clockwise

After the ECAP process, the surrounding Cu tube was removed and density measurements of the compacted samples were performed. The actual density of the sample (ρ_a) was measured using the Archimedes method, while the theoretical density (ρ_t) was estimated using the rule of mixture. The porosity (Φ) of the sample was estimated using the following equation [25]:

$$\Phi = 1 - \rho_a/\rho_z. \tag{1}$$

A compression test was performed at room temperature using a universal material testing machine MTS810 (MTS, Eden Prairie, MN, USA). Samples with a size of 7.5 mm × 5 mm × 5 mm were tested under a strain rate of 10^{-3}/s. Nickel anti-seize paste was used as a lubricant between the sample surface and anvil. Vickers microhardness tests were performed using Akashi MVK-H1(Mitutoyo, Kawasaki, Japan) microhardness tester. The load and the dwell time were 100 g and 15 s, respectively.

Microstructural characterization was conducted with an optical microscope (OM, OLYMPUS, Tokyo, Japan) and a field-emission scanning electron microscope (FE-SEM) (JEOL JSM-6500F, JEOL, Tokyo, Japan) equipped with energy dispersive spectroscopy (EDS, Oxford Instrument, Abingdon, UK). The particle size distribution of the as-milled powders was measured by a laser particle size analyzer (Malvern Mastersizer 2000, Malvern Panalytical, Malvern, UK). Phase analysis was performed by X-ray diffraction (XRD) (D2 PHASER, Brukcr, Madison, WI, USA) using Cu $K\alpha$ radiation. The scan range was from 20° to 90° and the data were collected with a step size of 0.02° and time of 0.5 s. The crystallite size of Mg was calculated from the broadening of the respective XRD peaks. Since the peak broadening is influenced by the crystallite size and microstrain, it is necessary to separate these two contributions. The separation of crystallite size and strain was obtained from Cauchy/Gaussian approximation using the following equation [26]:

$$\delta^2(2\theta)/\tan^2\theta = (k\lambda/L)(\delta(2\theta)/\tan\theta\cdot\sin\theta) + 16e^2 \tag{2}$$

$$\delta(2\theta) = B(1 - b^2/B^2) \tag{3}$$

where L is the mean crystallite size, k is constant (~1) and e is microstrain. λ is the wavelength and θ is the position of the analyzed peak maximum. The term $\delta(2\theta)$ is used to correct instrumental broadening. B and b are the breadths of the XRD peaks of the sample and the reference, respectively.

3. Results and Discussion

3.1. Microstructural Characterization

3.1.1. Mechanically-Alloyed Mg-Zn-Y Powders

SEM micrographs of $Mg_{97}Zn_1Y_2$ powders after mechanical alloying for 1, 10, 20 and 30 h are shown in Figure 1. As shown in the figure, the particle size of the powder decreased continuously when the milling time increased from 1 h to 20 h. After 20 h of milling, an increase of particle size was observed due to the agglomeration between powder particles. The same trend of the evolution of particle size was also obtained by measurement using a particle size analyzer (Figure 2). The particle size of the milled powder decreased from 100 μm to 28 μm when the milling time increased from 1 h to 20 h and then increased to 50 μm when the milling time reached 30 h. XRD-calculated crystallite sizes of Mg decreased from 59 nm to 36 nm when the milling time increased from 1 h to 20 h and then remained almost unchanged when the milling time reached 30 h. The evolution of particle size and crystallite size implied that fracturing and welding of the powders in the milling process had reached equilibrium after milling for 20 h. Thus, the 20-h-milled powder was selected for further compacting by the ECAP process.

XRD analysis was performed to investigate phase evolution during the mechanical alloying process. As shown in Figure 3a, intensities of Mg peaks decreased and became broad with the increase of milling time, indicating the refinement of the crystallite size of the Mg phase. In addition, Mg peaks gradually shifted to a higher diffraction angle. This indicated the lattice parameters of Mg decreased with the increase of milling time, which was confirmed by the calculated lattice parameters of Mg in the MA specimens (Table 2). The dissolution of Zn in Mg will increase the lattice parameters of Mg, while the dissolution of Y will do the opposite. The intensity of the Zn peak decreased as milling time increased, and the Zn peak disappeared after milling for 20 h. In contrast, the Y peak was still observed after milling for 20 h. The results of the XRD analysis indicated that most of Zn dissolved in Mg and only a small portion of Y dissolved after milling for 20 h. The dissolution of Zn was the main factor that resulted in a decrease of the lattice parameters of Mg. The formation of Y_2O_3 was observed after milling for 1 h. After milling for 20 h, peaks belonged to α-Mg, Y and Y_2O_3 phases were observed. No Mg-Zn-Y, Mg-Y or Mg-Zn intermetallic phases were detected after milling for 20 h. In contrast to the formation of $Mg_{24}Y_5$ after milling of a mixture of Mg, Zn and Y powder for 5 h in a SPEX mill, which was reported by Koch et al. [16], no Mg-Y intermetallic phases were synthesized in the milled powder in the present work. This was due to the reason that the planetary mill used in the present study had lower mechanical energy than that of the SPEX mill.

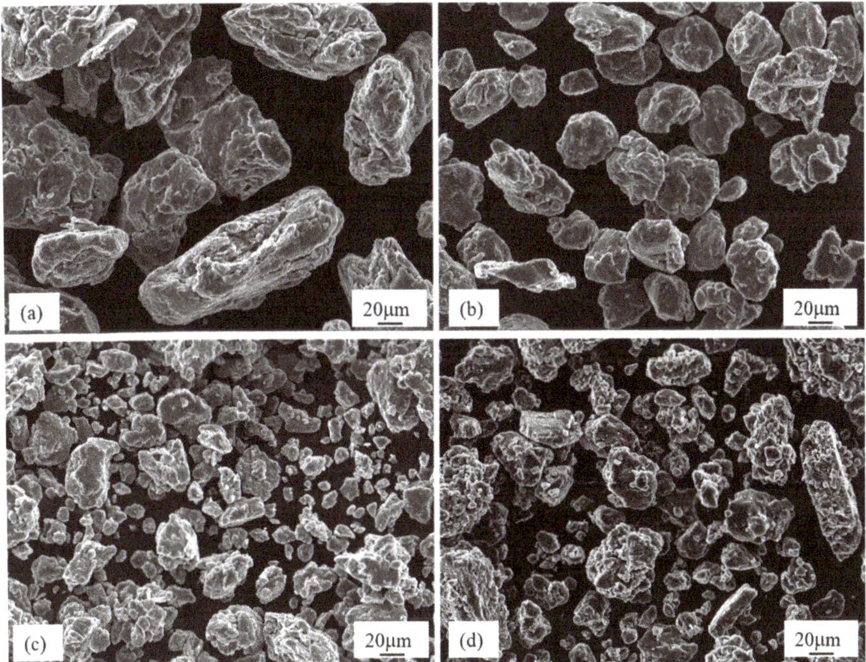

Figure 1. Morphology of powder mixture after mechanical alloying for (**a**) 1 h, (**b**) 10 h, (**c**) 20 h and (**d**) 30 h.

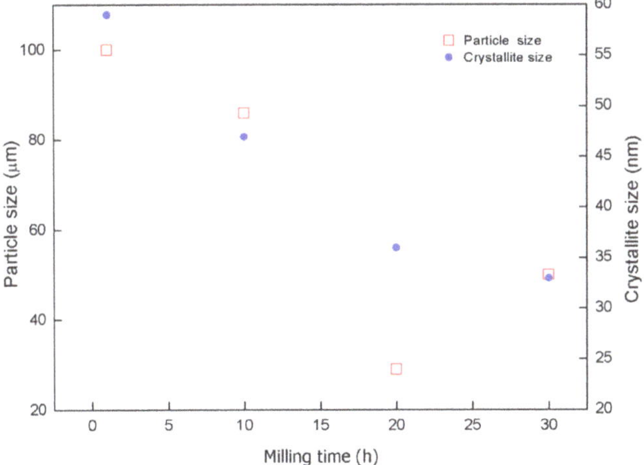

Figure 2. Evolution of particle size and crystallite size as a function of milling time.

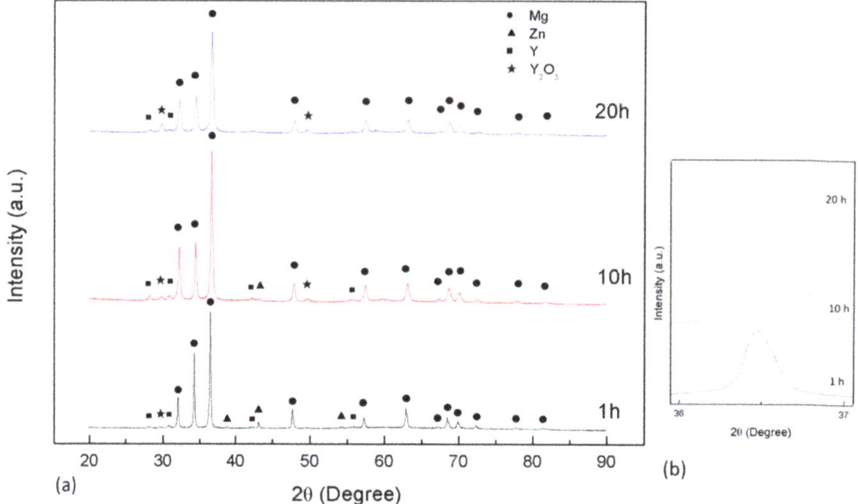

Figure 3. XRD patterns of (**a**) Mg$_{97}$Zn$_1$Y$_2$ alloy after milling for 1 h, 10 h, and 20 h; (**b**) pattern taken from 36° to 37° showing shifting of (101) Mg peak.

Table 2. Particle size, XRD-calculated crystallite size, lattice parameters and unit cell volume of Mg, porosity, and a fraction of oxide in the MA powders after different processing time.

Milling Time (h)	Particle Size (µm)	Crystallite Size (nm)	Lattice Parameter, a (nm)	Lattice Parameter, c (nm)	Unit Cell Volume (10^{-2}·nm^3)
1	100	59	0.3217	0.5223	4.6811
10	86	47	0.3211	0.5212	4.6528
20	29	36	0.3208	0.5208	4.6430
30	50	32	0.3208	0.5207	4.6417

3.1.2. Mg-Zn-Y Powders Consolidated by ECAP

XRD patterns of Mg$_{97}$Zn$_1$Y$_2$ powders after compacting with ECAP route Bc or C for one, two and four passes at 300 °C are shown in Figure 4. The pattern of 20-h-milled powder was also included for comparison. After one pass, strong Mg peaks were still observed. The Y peak, which was found in the 20-h-milled sample, disappeared and the intensity of the Y$_2$O$_3$ peak increased, indicating the oxidation of retained Y during the ECAP process. The formation of MgO was also observed after one pass. No shifting of Mg peaks was observed, implying the retained Y did not dissolve in Mg but oxidized to form Y$_2$O$_3$ during the ECAP process. A further increase in the number of passes or changing ECAP route to route C did not alter the phases presented in the compacted Mg$_{97}$Zn$_1$Y$_2$ samples. No phase transformation occurred when the samples underwent four passes. All of the samples compacted by ECAP route Bc or C contained α-Mg, Y$_2$O$_3$ and MgO phase. In the present work, the number of passes and route had no effect on the formation of phases in the consolidated bulk samples. There was no formation of binary Mg-Y and Mg-Zn or ternary Mg-Zn-Y phases after ECAP. During the MA process, part of the Y in the powder oxidized and formed Y$_2$O$_3$. In the following ECAP process, the retained Y reacted with oxygen and the formation of Y$_2$O$_3$ was completed. The results showed that an Mg alloy with dispersed Y$_2$O$_3$ particles was successfully synthesized during MA and the followed ECAP process.

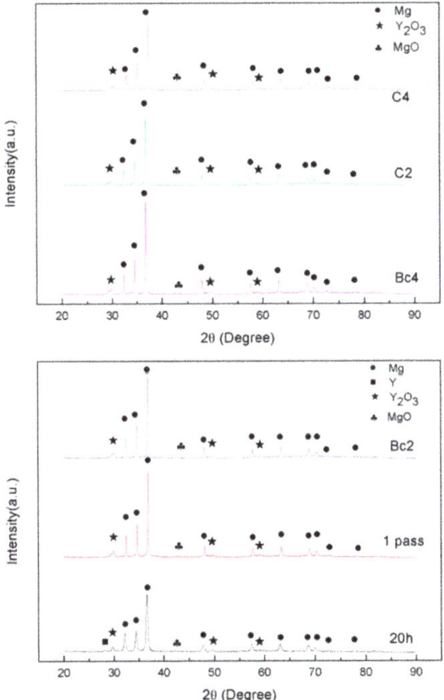

Figure 4. XRD patterns of powder after compacting by ECAP route C and Bc for one, two and four pass. The pattern of the 20-h-milled sample is also included for comparison.

Crystallite size of Mg increased from 36 nm in the 20-h-milled sample to 77 nm in the sample after one pass of ECAP route Bc and decreased to 64 nm when the number of passes increased to four. The observation that crystallite growth after one pass of ECAP at 300 °C and refinement after four passes agreed with the reported results that the ECAP process was able to refine the grain size when the number of passes increased [27]. The refinement of crystallite size was not observed in the samples processed with ECAP route C. The refinement seemed to be impeded by the oxide layer in the outer region of the sample.

Microstructures of the $Mg_{97}Zn_1Y_2$ powder samples consolidated by ECAP route Bc are shown in Figure 5. The one-pass sample consisted of three zones. EDS analysis results indicated that matrix (zone 1) was an Mg-rich phase (99.75 at% Mg, 0.24 at% Zn and 0.01 at% Y). In comparison to the matrix, dispersed particle (zone 2) was richer in Y and O (42.07 at% Mg, 31.12 at% Y and 26.81 at% O). A grey layer presented on the particle surfaces (zone 3) was an Mg-rich and O-rich phase (64.08 at% Mg and 35.93 at% O). Combining the results from the XRD and EDS analyses, it was concluded that the one-pass sample consisted of α-Mg (a solid solution containing a small amount of Zn and Y), dispersed Y_2O_3 particles and an MgO layer presented on the particle surfaces. After two passes, particles richer in Mg and O were identified in the Mg-matrix, indicating the oxide layer on the particle surface had been broken into particles. ECAP effectively broke down the MgO layer into small particles during the process. Compared to the two-pass sample, the composition of these three zones remained almost unchanged after four passes. This indicated that no phase transformation occurred after four passes. Porosities were also observed. The porosity of the one-pass, two-pass and four-pass samples was 9.7%, 7.6%, and 4.3%, respectively. The powder could be compacted more densely by the Bc route with the increasing number of passes.

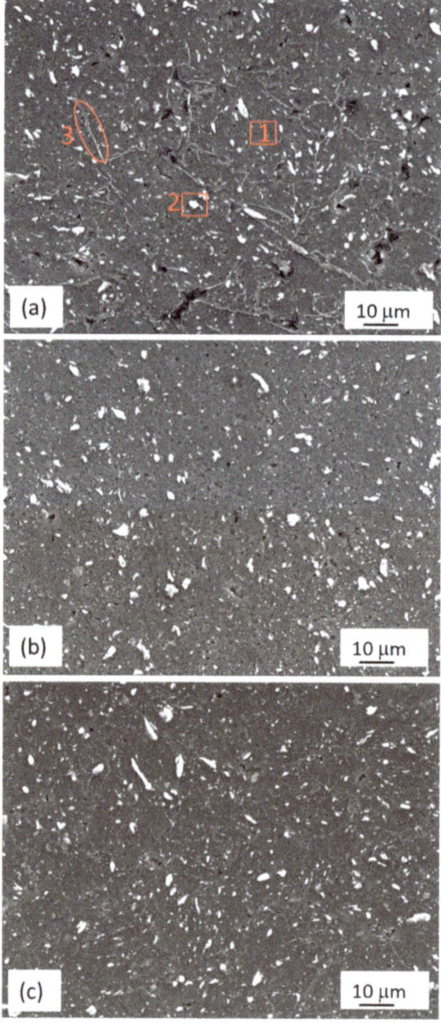

Figure 5. SEM micrograph of powder after compacting by ECAP route Bc for (**a**) one pass, (**b**) two passes and (**c**) four passes.

Figure 6a,b shows the microstructures of the powder samples pressed by route C for two and four passes, respectively. Similar to the powder compacted by route Bc, the samples pressed by route C also contained α-Mg, MgO and Y_2O_3 phases. However, the distribution of MgO was different in the inner and outer region in samples processed by route C. For the samples pressed by route C, MgO presented as dispersed particles in the α-Mg matrix in the inner portion of the sample (Figure 6a,b), while MgO was shown as a layer between Mg particles in the outer portion of the sample (Figure 6c). A clear boundary between the inner and outer zone could be observed (Figure 6d). The particles in the inner zone were compacted more effectively than those in the outer zone since the oxide layer has been broken into particles. The sample compacted by route C for four passes had a porosity of 8.1% while the sample compacted by route Bc for four passes had a porosity of 4.3%. The porosity of the sample pressed by route C was higher than that of the sample compacted by route Bc. No binary Mg-Zn, Mg-Y or ternary Mg-Zn-Y phases have been formed after the ECAP process.

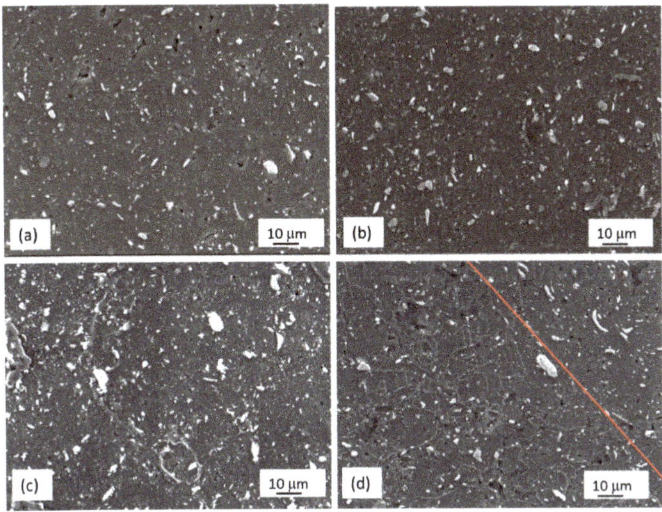

Figure 6. SEM micrograph of powder after compacting by ECAP route C for (**a**) two passes and (**b**) four passes; (**c**) outer zone of powder compacted by ECAP route C and (**d**) interface for interior and outer zone.

Shearing characteristics in an ECAP process can be discussed with shear patterns in three orthogonal planes of observation. The plane X was the mutually orthogonal plane of sectioning lying perpendicular to the longitudinal axis of the sample, while planes Y and Z were planes parallel either to the side faces or to the top face of the sample at the point of exit from the die, respectively [28]. Samples pressed using route C had deformation in the X and Y planes but no deformation in the Z plane. Samples compacted by route Bc had continuous deformation in three planes. Moreover, the strain path got reversed in the successive passes, enabling the easy formation of shear bands [29]. Thus, the samples pressed by route Bc had a higher compacting efficiency and lower porosity. The nonuniform distribution of shear force in the Z plane resulted in different morphology of MgO in the inner and outer region in the sample pressed by ECAP route C.

The fraction of oxide in the compacted samples was between 12.1% and 13.2% (Table 3), indicating the number of passes and type of route had no effect on the fraction of oxide. Most of the oxidation was completed after one pass of ECAP. No refinement of the crystallite size of Mg was observed in the present work. The porosity of the sample compacted by route Bc decreased from 9.7% after one pass to 4.3% after four passes (Table 3).

Table 3. XRD-calculated crystallite size of Mg, porosity and fraction of oxide in the ECAPed samples.

Route and Pass	Crystallite Size (nm)	Porosity (%)	Fraction of Oxide (%)
1 Pass	77	9.7	12.4
Bc 2	67	7.6	13.2
Bc 4	64	4.3	12.0
C2	73	9.2	12.1
C4	73	8.1	13.4

In the present work, the synthesis of Y_2O_3 reinforced Mg alloys during the MA and the subsequent ECAP process was achieved. The possible oxidation reactions and calculated Gibbs free energy of the reactions in a system with Mg, Y and O were as follows [30]:

$$2Mg + O_2 \rightarrow 2MgO, \Delta G = -1347 \text{ kJ/mol} \tag{4}$$

$$4Y + 3O_2 \rightarrow 2Y_2O_3, \Delta G = -4279 \text{ kJ/mol} \tag{5}$$

$$2Y + 3MgO \rightarrow 3Mg + Y_2O_3, \Delta G = -99 \text{ kJ/mol.} \tag{6}$$

According to Reactions (4) and (5), the formation motivation of Y_2O_3 was quite higher than that of MgO. Thus, the oxidation of Y in the Mg matrix was possible during the MA and the subsequent ECAP process. It was also possible that Y_2O_3 formed via Reaction (6) during the process.

3.2. Mechanical Properties

Table 4 lists Vickers microhardness of $Mg_{97}Zn_1Y_2$ powder compacted by ECAP route Bc or C at 300 °C. The Vickers hardness was in the range of 100 to 110 Hv. The type of route and number of passes did not show a strong effect on the hardness of compacted samples. Pure Mg powder after compacting at 300 °C by ECAP using route Bc for four times exhibited a hardness of 49 Hv [9]. As-cast $Mg_{97}Zn_1Y_2$ had a hardness of ~80 Hv [24]. As shown in Table 4, hardness values of the $Mg_{97}Zn_1Y_2$ powders compacted by ECAP at the present work were higher than those of compacted Mg powder and as-cast $Mg_{97}Zn_1Y_2$ ingot. The increase of hardness was due to the solid solution of Zn and Y, and the dispersion of oxides; the latter was the dominant factor.

Table 4. Vickers hardness, compressive strength and failure strain of ECAPed samples.

Route and Pass	Hardness (HV)	Compressive Strength (MPa)	Failure Strain (%)
1 Pass	101	67	8
Bc 2	101	90	3
Bc 4	110	185	4
C2	103	60	6
C4	100	44	1

The results of compressive tests for $Mg_{97}Zn_1Y_2$ compacted by ECAP route Bc and C for one, two, and four passes at 300 °C are summarized in Table 4. After one pass, the ultimate compressive strength (UCS) of the compacted sample reached 67 MPa and failure strain was 8%. After two passes of route Bc, UCS increased to 90 MPa, and failure strain decreased to 3%. A further increase in the pass number to four passes resulted in an increase in UCS (185 MPa). In contrast, UCS and failure strain of the sample after two passes of route C remained close to those of the one-pass sample. Moreover, those values even decreased after four passes of route C. UCS of the four-pass sample was lower than that of the pure Mg sample (190 MPa).

The strength of the compacted samples could be affected by the grain size of the matrix, particle size and fraction of dispersed phase, and porosity. In the present study, the fraction and size of the oxide phase remained almost unchanged for the compacted samples produced by ECAP routes C and Bc for one, two and four passes (Table 3). Therefore, they were not the dominant factors affecting the mechanical properties of the compacted samples. The XRD-calculated crystallite size and porosity decreased with the increasing number of passes in samples processed with route Bc. However, they remained almost the same in samples processed with route C. The sample pressed with route Bc for four passes possessed both of the finest crystallite size and the lowest porosity. Thus, the UCS of the sample after four ECAP passes with route Bc was enhanced but the failure strain was sacrificed as compared to the one-pass sample. Moreover, the distribution of the oxide (MgO) layer at the outer region of samples processed with route C also resulted in a lower UCS than those of samples compacted with route Bc. Unlike the dispersed oxide particles in the inner portion of the sample, these oxide layers lowered the bonding of the powders.

Compared to the UCS of compacted Mg samples reported by Lee et al. using the same ECAP parameters [9], the UCS of the samples in the present work was not enhanced even with the dispersion of Y_2O_3. This was due to the higher amount of porosity in the present work (4.3% in the present work in comparison to 1.6% in the pure Mg sample [9]). However, the hardness value was higher than that of the pure Mg sample. The reason was that Vickers hardness was measured in a small area with a minimal effect of porosity. On the other hand, the results obtained from the compression test were affected strongly by the presence of porosity.

Feasibility of simultaneous synthesis of dispersed Y_2O_3 particles and compaction of Mg alloy powders during MA and followed ECAP process was confirmed in the present study. However, the processing parameters needed to be optimized to improve the mechanical properties of the compacted Mg alloy samples in future work. Mallick et al. studied the deformation behavior of Mg-$2vol.\%Y_2O_3$ nanocomposite [31]. The composite was fabricated using the powder metallurgy (PM) method, including powder blending of pure Mg and Y_2O_3 powder (particle size of 30–50 nm), cold compaction, sintering and hot extrusion. The extruded sample had a strength of 291 MPa, which was higher than ours. In addition to decreasing the high porosity, the particle size of Y_2O_3 should be reduced and the fraction of Y_2O_3 needs to be optimized to improve the strength of the compacted samples.

4. Conclusions

In the present study, a bulk $Mg_{97}Zn_1Y_2$ alloy reinforced by Y_2O_3 particles was prepared by simultaneous synthesizing Y_2O_3 particles and compacting mechanically-alloyed powders using equal channel angular pressing. The main conclusions can be summarized as follows:

1. The $Mg_{97}Zn_1Y_2$ alloy powder after mechanical alloying for 20 h consisted of α-Mg, Y and Y_2O_3 phases. The ECAP-compacted bulk alloy contained the α-Mg matrix, uniformly dispersed Y_2O_3 and MgO phase.
2. The powder compacted with ECAP route Bc for four passes exhibited a hardness of 110 HV and an ultimate compressive strength of 185 MPa. The hardness was higher than those of as-cast $Mg_{97}Zn_1Y_2$ alloy and compacted Mg; however, the UCS was lower.
3. The improved hardness observed in the ECAP-compacted alloy was mainly attributed to the dispersion hardening of Y_2O_3 particles. The decrease in ultimate compressive strength was due to the higher amount of porosity in the ECAP-compacted alloy.

Author Contributions: Conceptualization, K.-L.O. and C.C.; Validation, C.C.; Formal Analysis, C.-C.C. and C.C.; Investigation, K.-L.O., C.-C.C. and C.C.; Writing-Original Draft Preparation, C.C.; Funding Acquisition, K.-L.O. All authors have read and agreed to the published version of the manuscript.

Funding: This research was funded by the Taipei Medical University and National Taiwan University of Science and Technology in Taiwan under TMU-NTUST joint research program contract no. TMU-NTUST-104-11.

Acknowledgments: The authors would like to thank Sheng-Chuan Liao (PIC, NTUST) for the technical support on SEM-EDS analysis.

Conflicts of Interest: The authors declare no conflict of interest.

References

1. Mordike, B.L.; Ebert, T. Magnesium: Properties-applications-potential. *Mater. Sci. Eng. A* **2001**, *302*, 37–45. [CrossRef]
2. Schumann, S. The paths and strategies for increased magnesium applications in vehicles. *Mater. Sci. Forum* **2005**, *488*, 1–8. [CrossRef]
3. Cabibbo, M.; Spigarelli, S. A TEM quantitative evolution of strengthening in an Mg-RE alloy reinforced with SiC. *Mater. Charact.* **2011**, *62*, 959–969. [CrossRef]

4. Pan, F.S.; Yang, M.B.; Chen, X.H. A review on casting magnesium alloys: Modification of commercial alloys and development of new alloys. *J. Mater. Sci. Technol.* **2016**, *32*, 1211–1221. [CrossRef]

5. You, S.; Huang, Y.; Kainer, K.U.; Hort, N. Recent research and developments on wrought magnesium alloys. *J. Magnesium Alloys* **2017**, *5*, 239–253. [CrossRef]

6. Figueiredo, R.B.; Langdon, T.G. Processing Magnesium and Its Alloys by High-Pressure Torsion: An Overview. *Adv. Eng. Mater.* **2019**, *21*, 1801039. [CrossRef]

7. Kawamura, Y.; Hayashi, K.; Inoue, A.; Masumoto, T. Rapidly solidified powder metallurgy $Mg_{97}Zn_1Y_2$ alloy with excellent tensile yield strength above 600 MPa. *Mater. Trans.* **2001**, *42*, 1172–1176. [CrossRef]

8. Inoue, A.; Matsushita, M.; Kawamura, Y.; Amiya, K.; Hayashi, K.; Koike, J. Novel hexagonal structure of ultra-high strength magnesium-based alloys. *Mater. Trans.* **2002**, *43*, 580–584. [CrossRef]

9. Lee, H.C.; Chao, C.G.; Liu, T.F.; Lin, C.Y.; Wang, H.C. Effect of temperature and extrusion pass on the consolidation of magnesium powders using equal channel angular extrusion. *Mater. Trans.* **2013**, *54*, 765–768. [CrossRef]

10. Li, G.; Sun, J.; Guo, Q.; Wang, R. Fabrication of the nanometer Al_2O_3/Cu composite by internal oxidation. *J. Mater. Process. Tech.* **2005**, *170*, 336–340. [CrossRef]

11. Broyles, S.E.; Anderson, K.R.; Groza, J.R.; Gibeling, J.C. Creep deformation of dispersion-strengthened copper. *Metall. Mater. Trans A* **1996**, *27*, 1217–1227. [CrossRef]

12. Rapp, R.A. Kinetics, Microstructures and mechanism of internal oxidation—Its effect and prevention in high temperature alloy oxidation. *Corrosion* **1965**, *21*, 382–401. [CrossRef]

13. Chino, Y.; Furuta, T.; Hakamada, M.; Mabuchi, M. Influence of distribution of oxide contaminants on fatigue behavior in AZ31 Mg alloy recycled by solid-state processing. *Mater. Sci. Eng. A* **2006**, *424*, 355–360. [CrossRef]

14. Suryanarayana, C. Mechanical alloying and milling. *Prog. Mater. Sci.* **2001**, *46*, 1–184. [CrossRef]

15. Zhou, H.; Hu, L.; Sum, Y.; Zhang, H.; Duan, C.; Yu, H. Synthesis of nanocrystalline AZ31 magnesium alloy with titanium addition by mechanical milling. *Mater. Charact.* **2016**, *113*, 108–116. [CrossRef]

16. Koch, C.C.; Scattergood, R.O.; Youssef, K.M.; Chan, E.; Zhu, Y.T. Nanostructured materials by mechanical alloying: New results on property enhancement. *J. Mater. Sci.* **2010**, *45*, 4725–4732. [CrossRef]

17. Senkov, O.N.; Miracle, D.B.; Scott, J.M.; Senkova, S.V. Compaction of amorphous aluminum alloy powder by direct extrusion and equal angular extrusion. *Mater. Sci. Eng. A* **2005**, *393*, 12–21. [CrossRef]

18. Kim, H.S.; Seo, M.H.; Oh, C.S.; Kim, S.J. Equal channel pressing of metallic powders. *Mater. Sci. Forum* **2003**, *437*, 89–92. [CrossRef]

19. Nagasekhar, A.V.; Yip, T.H.; Ramakanth, K.S. Mechanics of single pass equal channel angular extrusion pf powder in tubes. *Appl. Phys. A* **2006**, *85*, 185–194. [CrossRef]

20. Baker, I.; Iliescu, D.; Liao, Y. Containerless consolidation of Mg powders using ECAE. *Mater. Manuf. Processes* **2010**, *25*, 1381–1384. [CrossRef]

21. Nagasekhar, A.V.; Yip, T.H.; Guduru, R.K.; Ramakanth, K.S. Multipass equal channel angular pressing of MgB_2 powder in tubes. *Physica C* **2007**, *466*, 174–180. [CrossRef]

22. Karaman, I.; Haouaoui, M.; Maier, H.J. Nanoparticle consolidation using equal channel angular extrusion at room temperature. *J. Mater. Sci.* **2007**, *42*, 1561–1576. [CrossRef]

23. Chiu, C.; Lu, C.; Chen, S.; Ou, K. Effect of Hydroxyapatite on the Mechanical Properties and Corrosion Behavior of Mg-Zn-Y Alloy. *Materials* **2017**, *10*, 885. [CrossRef] [PubMed]

24. Chiu, C.; Huang, H. Microstructure and properties of Mg-Zn-Y alloy powder compacted by equal channel angular pressing. *Materials* **2018**, *11*, 1678. [CrossRef] [PubMed]

25. Ramya, M.; Sarwat, S.G.; Udhayabanu, V.; Subramanian, S.; Raj, B.; Ravi, K.R. Role of partially amorphous structure and alloying elements on the corrosion behavior of Mg–Zn–Ca bulk metallic glass for biomedical applications. *Mater. Des.* **2015**, *86*, 829–835. [CrossRef]

26. Klug, H.P.; Alexander, L. *X-ray Diffraction Procedures for Polycrystalline and Amorphous Materials*, 2nd ed.; John Wiley & Sons: New York, NY, USA, 1974; pp. 618–708.

27. Figueiredo, R.B.; Langdon, T.G. Grain refinement and mechanical behavior of a magnesium alloy processed by ECAP. *J. Mater. Sci.* **2010**, *45*, 4827–4836. [CrossRef]

28. Furukawa, M.; Iwahashi, Y.; Horita, Z.; Nemoto, M.; Langdon, T.G. The shearing characteristics associated with equal-channel angular pressing. *Mater. Sci. Eng. A* **1998**, *257*, 328–332. [CrossRef]

29. Iwahashi, Y.; Horita, Z.; Nemoto, M.; Langdon, T.G. An investigation of microstructural evolution during equal-channel angular pressing. *Acta Mater.* **1997**, *45*, 4733–4741. [CrossRef]
30. Barin, I. *Thermochemical Data of Pure Substance*, 3rd ed.; VCH Publishers: New York, NY, USA, 1995; p. 993.
31. Mallick, A.; Tun, K.S.; Gupta, M. Deformation behavior of Mg/Y_2O_3 nanocomposite at elevated temperatures. *Mater. Sci. Eng. A* **2012**, *551*, 222–230. [CrossRef]

Communication

Analysis of Nanoprecipitation Effect on Toughness Behavior in Warm Worked AA7050 Alloy

Claudio Testani [1,*], Giuseppe Barbieri [2] and Andrea Di Schino [3]

[1] CALEF, Management Department, c/o ENEA CR Casaccia, Via Anguillarese 301, Santa Maria di Galeria, 00123 Rome, Italy

[2] MATPRO Department, ENEA CR Casaccia, Via Anguillarese 301, Santa Maria di Galeria, 00123 Rome, Italy; Giuseppe.barbieri@enea.it

[3] Department of Engineering, University of Perugia, Via G. Duranti 93, 06125 Perugia, Italy; andrea.dischino@unipg.it

* Correspondence: claudio.testani@consorziocalef.it; Tel.: 1 39-06-3048-4653

Received: 23 November 2020; Accepted: 17 December 2020; Published: 20 December 2020

Abstract: Commonly adopted main methods aimed to improve the strength–toughness combination of high strength aluminum alloys are based on a standard process. Such a process includes alloy solution heat treatment, water-quench and reheating at controlled temperature for ageing holding times. Some alloys request an intermediate cold working hardening step before ageing for an optimum strength result. Recently a warm working step has been proposed and applied. This replaces the cold working after solution treatment and quenching and before the final ageing treatment. Such an alternative process proved to be very effective in improving strength–toughness behavior of 7XXX aluminum alloys. In this paper the precipitation state following this promising process is analyzed and compared to that of the standard route. The results put in evidence the differences in nanoprecipitation densities that are claimed to be responsible for strength and toughness improved properties.

Keywords: AA7050 alloy; warm working; toughness

---

1. Introduction

Lightweight metals and alloys represented for many years the most suitable solution for many high-tech applications, including sport equipment [1], energy and automotive [2]. Aerospace has probably been the sector where most of the potential of aluminum and titanium resides. Among lightweight metals, aluminum alloy are gaining huge industrial significance because of their outstanding combination of mechanical, physical and tribological properties [3]. Following such behavior, they appear to be the best candidates for structural aerospace designers if compared with other alloys [4–6]. The origin of their peculiar behavior, depending on different alloying strategies and processes, has been investigated by many researchers in the past years [7,8].

Alloying elements are selected based on their individual properties as they impact on microstructure and performance characteristics. In this framework, the effects of the main alloying elements in aluminum alloys have been reported in detail by Mondolfo in [9]; Sauvage et al. [10] report about the influence of Cu in hardening behavior of ultrafine-grained Al-Mg-Si alloys.

Among aluminum alloys, AA7050 is one of the best performing taking into account the good balance of high strength, corrosion resistance and toughness, making it largely adopted in aerospace applications [11,12]. Such properties are achieved by recrystallization phenomena during and after hot forming processes [13–16]. Wang et al. recently reported about a physically based constitutive analysis and microstructural evolution investigation in AA7050 aluminum alloy during hot compression [17]. Maizza et al. proposed a recrystallization model for aluminum alloys [18] to be applied to components with complex geometric shape, often manufactured by closed-die forging followed by alloy solution heat treatment followed by water-quench and reheating at controlled temperature for ageing holding times. MacKenzie reports in detail about heat treatment effect aluminum forged components [19].

AMS 4333 International Standard prescribes for such alloys an intermediate cold deformation step (max 5% of cold upsetting) between the water quench, after solution treatment and before the two-stage ageing final step, aimed to reach the optimal precipitate distribution and guarantee the best mechanical properties [20]. One alternative process has been originally proposed by Wyss in the United States Patent US5194102A [21], reporting about the beneficial effect on fracture toughness exerted by an intermediate warm hardening process replacing the cold upsetting step. Such a route has been proved to be successful in 7xxx alloys leading to an improvement in toughness behavior without any detrimental effects in terms of hardness and tensile properties [22]. Moreover, such a process reduces the microstructural heterogeneity by grain refining [23]. As a matter of fact, dislocation cross-slip occurring during deformation in this temperature range allows a general grain reorientation and the sub-boundaries present inside the grains tend to evolve towards high-angle boundaries: the higher the deformation temperature, the easier the process [24]. The process, reported in many papers [25–27], is defined by some authors as continuous dynamic recrystallization (CDR). Other phenomena, such us strain hardening [28,29] and recovery [30], depend on dislocation structure evolution and on its interaction with other crystallographic defects. Cross-slip is recognized as an effective mechanism because it allows dislocations to bypass obstacles and affects the material's final microstructure [31].

In this paper the precipitation state of three different process routes (one standard) are analyzed in correlation with toughness behavior.

2. Materials and Experimental Details

The AA7025 alloy nominal chemical composition is reported in Table 1.

Table 1. Nominal chemical composition of the AA7050 alloy (wt %).

Elements	Al	Zr	Si	Fe	Cu	Mn	Mg	Cr	Zn	Ti
wt (%)	Bal.	0.12	<0.12	<0.15	2.3	<0.1	2.2	<0.04	6.25	<0.05

Three families of squared 10 cm × 6 cm × 3 cm samples (namely family A, (samples A1, A2 and A3), family B (samples B1, B2 and B3) and family C (samples C1, C2 and C3)) were machined starting from a round bar, with a starting diameter $D = 120$ mm. The bar was hot forged at T > 400 °C with 75% total deformation. The heat treatment was adopted in agreement with the requirements from the standard AMS2770N specification since the fulfillment of this standard is mandatory for AA7050 alloy aeronautical forged component manufacturers. The process was completed with room temperature upsetting and final two stages aging at 5 h at 394 K + 8 h at 450 K (specimens A). The two innovative cycles (specimens B and C) just varied from AMS2770N specification requirements in terms of upsetting temperature: in fact, they were carried out at 423 K and 473 K instead of room temperature, while all the other cycle steps were unmodified (Table 2). The two temperatures are close and straddling the second ageing step temperature (T = 450 K). Moreover, in the literature [21] it is shown that upsetting temperatures higher than 473 K can affect hardness values.

Table 2. Modified heat treatment cycles for the AA7050 alloy.

Sample Family	Solution Heat Treatment T = 748 K for 5 h	Water Quenching	Upsetting Temperature (Deformation Max 5%)	First Ageing Step: 394 K for 5 h	Second Ageing Step: 450 K for 8 h
A (1–3)	YES	YES	293 K	YES	YES
B (1–3)	YES	YES	423 K	YES	YES
C (1–3)	YES	YES	473 K	YES	YES

All samples were subjected to solution heat treatment at T = 748 K for 5 h, water quenching, 5% warm deformation and ageing. The process conditions differ on the upsetting temperature, according to Table 2. A1, A2 and A3 specimens were deformed at room temperature; samples B1, B2 and B3 at 423 K and samples; C1, C2 and C3 at 473 K. After heat treatment transverse specimens (in agreement with ASTM-E399) were machined and tested for plain strain fracture toughness (KIC) tests on the transverse specimens (according to the ASTM E399 standard). Samples machined starting from the three families A, B and C were prepared for metallographic examination. Grain size was measured by light microscopy (LM) according to the ASTM E112 specification. Precipitation state analysis was performed with a field emission gun scanning electron microscope (FEG-SEM; SEM FEG LEO 1550 ZEISS equipped with an EDS OXFORD X ACT system, (version v2.2, Oxford Instruments NanoAnalysis & Asylum Research, Abingdon, UK). Ten micrographs have been examined for each sample in order to establish a statistical base for the quantitative precipitates' analysis. Precipitates number counting has been performed by means of IMAGE-J Fiji, (version 1 46, National Institutes of Health, Bethesda, MD, USA), a software for the automatic images processing and analyses program.

An example of SEM FEG image prepared for precipitation number count by mean of IMAGE-J Fiji 1.46 software is reported in Figure 1.

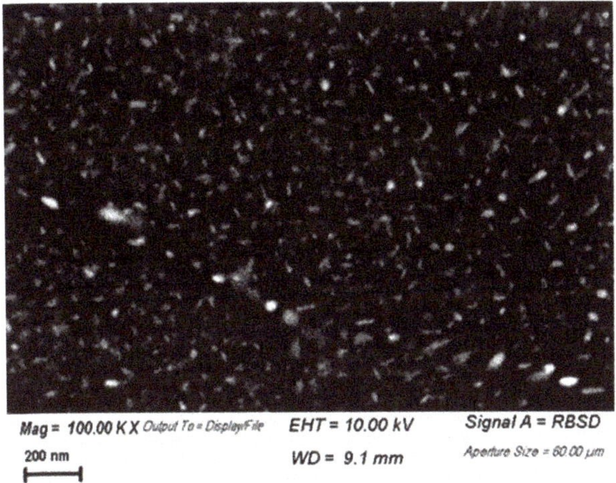

Figure 1. Example of the SEM-FEG high magnification (100 KX) image in sample A.

3. Results and Discussion

Microstructural investigation highlighted an actual grain size (D_g) refinement with the increasing upsetting temperature (Figure 2). At the same time, mechanical assessment results (as reported in detail in [22,23] and summarized in Table 3 with the results of averaged grain size) show that, even if hardness and yield strength showed a slight increase, K_{IC} performance showed a not negligible improvement. As a matter of fact, the innovative process with upsetting at T = 473 K led to an enhancement of about 10% in fracture toughness K_{IC} in comparison to the process cycle carried on in agreement with the AMS2770N specification.

It can be observed that both hardness and tensile mechanical properties increased following the additional upsetting temperature. This suggests that a significant change in fine precipitation occurred, mainly formed from Guiner Preston zones [32,33], acting in terms of grain refinement. Such investigation, aimed to support the mechanism underlying the grain refinement with warm upsetting temperature increase, was not previously carried out, and it is discussed below.

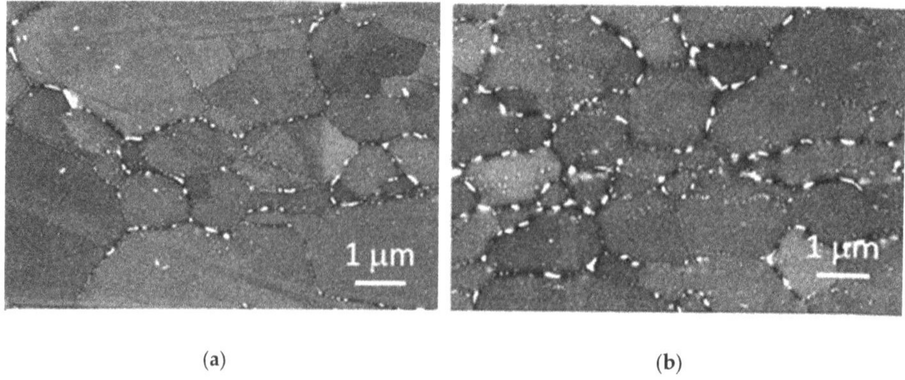

(a)   (b)

Figure 2. Warm upsetting temperature effect on the AA7050 alloy microstructure. (**a**) Warm upsetting temperature = 293 K, average D_g = 10.1 μm and (**b**) warm upsetting temperature = 473 K, average D_g = 7.5 μm (SEM-FEG images at 20 KX magnification).

Table 3. Mechanical properties and mean grain size dependence on warm upsetting temperature in Alloy AA7050.

Sample	Warm Upsetting Temperature	Brinell Hardness, HB	Yield Strength, YS (MPa)	K_{IC} Mean Value (MPa m$^{1/2}$)	Average Grain Size (D_g Microns)
Sample A (three tests)	293 K	145	450	26.8	10.1
Sample B (three tests)	423 K	147	455	28.3	8.6
Sample C (three tests)	473 K	152	470	30.1	7.5

In order to measure the possible evolution of the overall precipitation status, digital image analysis was performed by setting a Feret diameter threshold of 10 nm. As an example SEM-FEG images of specimens A, B and C are reported in Figure 3.

The scanning electronic microscope SEM-FEG examination of the A, B and C specimens allowed us to have evidence of the following precipitate groups with respect to the size:

- Larger particles at the grain and subgrain boundaries (size ranging = 100–500 nm);
- Fine particles inside grains subgrains (size ranging = 20–100 nm);
- Very fine particles inside grains subgrains (size ranging ≤ 20 nm).

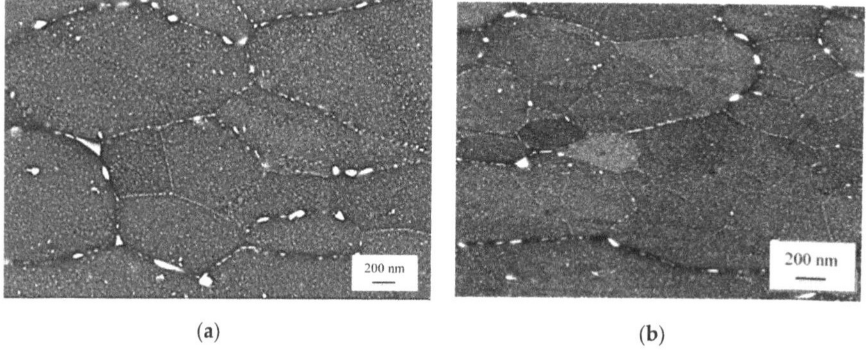

(a)   (b)

Figure 3. *Cont.*

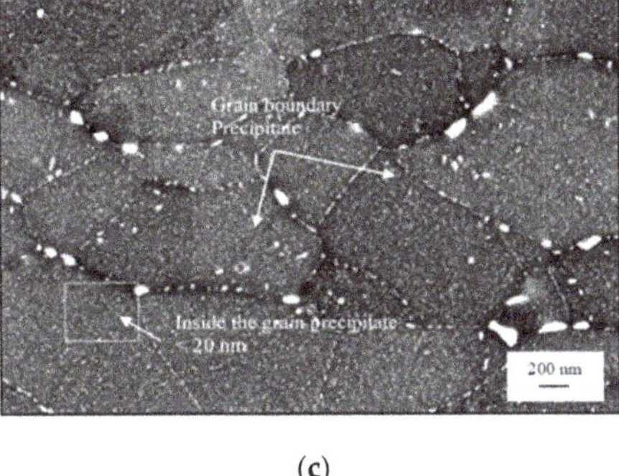

(**c**)

Figure 3. Warm upsetting temperature effect on the precipitation state of the AA7050 alloy. (**a**) Specimen A (upsetting at T = 293 K); (**b**) specimen B (upsetting at T = 423 K) and (**c**) specimen C (upsetting at T = 473 K; SEM-FEG images at 50 KX magnification).

In order to carry out an accurate precipitation population assessment, SEM-FEG images at 50 KX and 100 KX were selected. It must be reported that it was possible to detect some larger and isolated particles up to 5 µm at grain boundaries, which revealed to be very useful for the chemical analyses [22].

The number of detected precipitates as a function of intermediate upsetting temperature is reported in Figure 4, grouped for different precipitate size ranges.

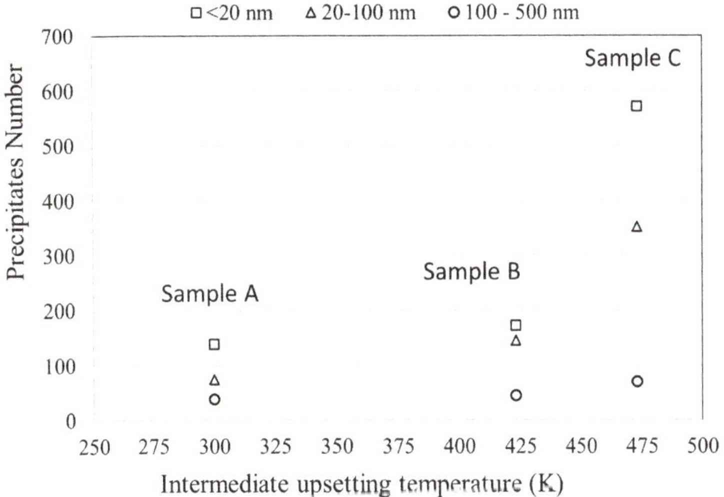

Figure 4. Mean values of precipitates number grouped in size classes (sample A: intermediate upsetting temperature = 293 K; sample B: intermediate upsetting temperature = 423 K; sample C intermediate upsetting temperature = 473 K).

Figure 4 clearly shows that the mean number of precipitates increased, and the intermediate upsetting deformation temperature increased, in the temperature range from 293 to 473 K.

In particular, the mean precipitates number as measured after upsetting at room temperature was very similar to the measured one after warm deformation at 423 K.

An evident increase of the finest precipitates (<20 nm) number was found after upsetting at 473 K. Moreover, the increase in upsetting temperature did not appear to affect the largest precipitates. This observation is in agreement with the detected improved toughness and yield strength behavior.

The above results suggest that the appearance of a fine and homogeneous precipitation state will act as a pinning effect at grain boundaries, according to the Zener effect [34]: this will favor a grain refinement down to 7.5 μm after upsetting at 473 K, as reported in Table 2. Such an effect will allow an increase in terms of tensile yield strength together with an improvement in terms of fracture propagation resistance according to the Griffith model [35]. Such a model dictates that the critical stage for cracking is the propagation of a small crack originated in a single grain to the adjacent one and is therefore opposed by grain refinement.

4. Conclusions

Results related to the AA 7050 alloy after a modified process route including a warm deformation stage show that:

- The AA7050 alloy precipitation state was very sensitive to heat treatments and especially to the intermediate deformation step included in the heat treatment cycle;
- The warm deformation temperature of 473 K resulted in the highest nanosized particles precipitation, resulting in increased deformation induced precipitation phenomenon;
- The higher nanosized mean value of precipitations obtained with warm intermediate upsetting steps claimed to be responsible of a grain refinement effect;
- The finer and more homogeneous microstructure resulted in a toughness K_{IC} properties improvement with respect to those obtained with room temperature intermediate upsetting step;
- The mechanism underlying the observed behavior was expected to be explained based on the Griffith model for crack propagation, stating that grain boundaries density increased, following a pinning effect. This phenomenon was favored by the presence of fine precipitates and will oppose crack propagation thus improving fracture toughness behavior.

Author Contributions: Conceptualization, C.T. and A.D.S.; Material preparation, A.D.S. and G.B.; SEM-FEG tests; SEM observations G.B. and C.T.; Writing C.T, G.B. and A.D.S. All the authors discussed the results. All authors have read and agreed to the published version of the manuscript.

Funding: This research received no external funding.

Conflicts of Interest: The authors declare no conflict of interest.

References

1. Napoli, G.; Di Schino, A.; Paura, M.; Vela, T. Colouring titanium alloys by anodic oxidation. *Metalurgija* **2018**, *57*, 111–113.
2. Sharma, D.K.; Filipponi, M.; Di Schino, A.; Rossi, F.; Castaldi, J. Corrosion behavior of high temperature fuel cells: Issues for materials selection. *Metalurgija* **2019**, *58*, 347–351.
3. Warren, A.S. Developments and challenges for aluminium—A Boeing perspective. *Mater. Forum* **2004**, *28*, 24–31.
4. Gloria, A.; Montanari, R.; Richetta, M.; Varone, A. Alloys for Aeronautic Applications: State of the Art and Perspectives. *Metals* **2019**, *9*, 662. [CrossRef]
5. *Aerospace Structural Materials Handbook*; DoD, Wright-Patterson Air Force Base: Dayton, OH, USA, 2001.
6. Properties and Selection: Nonferrous Alloys and Special Purpose Materials. In *ASM Handbook*; American Society of Materials: Cleveland, OH, USA, 2005; Volume 2.
7. Fayomi, O.S.; Popoola, P.; Udoye, N. Effect of Alloying Element on the integrity and Functionality of aluminium-based alloys. In *Aluminium Alloys—Recent Trends in Processing, Characterization, Mechanical Behavior and Applications*; IntechOpen: London, UK, 2017. [CrossRef]
8. Ranam, R.S.; Purohit, R.; Das, S. Reviews on the Influences of Alloying elements on the Mmcrostructure and Mechanical Properties of Aluminum Alloys and Aluminum Alloy Composites. *Int. J. Sci. Res.* **2012**, *2*, 1–7.

9. Mondolfo, L.F. *Aluminum Alloys: Structure and Properties*; Butterworth: London, UK, 1976; pp. 497–499.

10. Sauvage, X.; Lee, S.; Matsuda, K.; Horita, M. Origin of the influence of Cu or Ag micro-additions on the age hardening behavior of ultrafine-grained Al-Mg-Si alloys. *J. Alloys Compd.* **2017**, *710*, 199–202. [CrossRef]

11. Sanchez, J.M.; Rubio, E.; Alvarez, M.; Sebastian, M.A.; Marcos, M. Microstructural characterization of material adhered over cutting tool in the dry machining of aerospace aluminium alloys. *J. Mater. Process. Technol.* **2005**, *164–165*, 911–918. [CrossRef]

12. Rambabu, P.; Prasad, N.E.; Kutumbarao, V.V.; Wanhill, R.J. Aluminium Alloys for Aerospace Applications. In *Aerospace Materials and Material Technologies*; Springer: Singapore, 2017; Volume 1, pp. 9–52.

13. Testani, C.; Ielpo, F.M.; Alunni, E. AA2618 and AA7075 alloys superplastic transition in isothermal hot-deformation tests. *Mater. Des.* **2001**, *21*, 305–310. [CrossRef]

14. Zhao, Y.H.; Liao, X.Z.; Hin, Z.; Valiev, R.Z. Microstructures and mechanical properties of ultrafine grained 7075 Al alloy processed by ECAP and their evolutions during annealing. *Acta Mater.* **2004**, *52*, 4589–4599. [CrossRef]

15. Rufini, R.; Di Pietro, O.; Di Schino, A. Predictive Simulation of Plastic Processing of Welded Stainless Steel Pipes. *Metals* **2018**, *8*, 519. [CrossRef]

16. Di Schino, A.; Di Nunzio, P.E.; Turconi, G.L. Microstructure evolution during tempering of martensite in medium carbon steel. *Mater. Sci. Forum* **2007**, *558–559*, 1435–1441. [CrossRef]

17. Wang, S.; Luo, J.; Hou, L.; Zhang, J.; Zhuang, L. Physically based constitutive analysis and microstructural evolution of AA7050 aluminum alloy during hot compression. *Mater. Des.* **2016**, *107*, 277–289. [CrossRef]

18. Maizza, G.; Pero, R.; Richetta, M.; Montanari, R. Continuous dynamic recrystallization (CDRX) model for aluminum alloys. *J. Mater. Sci.* **2017**, *53*, 4563–4573. [CrossRef]

19. MacKenzie, D. Scott Heat Treating Aluminum for Aerospace Applications. *Heat Treat. Prog.* **2005**, *5*, 37–43.

20. *Aluminum Alloy, Die Forgings, 6.2Zn–2.3Cu–2.2Mg–0.12Zr (7050-T7452), Solution Heat Treated, Compression Stress-Relieved, and Overaged*; AMS 4333 International Standard; SAE International: Warrendale, PA, USA, 2015.

21. Wyss, R. ALCOA, Method for Increasing the Strength of Aluminum Alloy Products through Warm Working. U.S. Patent 5,194,102, 16 March 1993.

22. Angella, G.; Di Schino, A.; Donnini, R.; Richetta, M.; Testani, C.; Varone, A. AA70323 Al alloy hot-forging process for improved fracture toughness properties. *Metals* **2019**, *9*, 64. [CrossRef]

23. Tedde, M.; Di Schino, A.; Donnini, R.; Montanari, R.; Richetta, M.; Santo, L.; Testani, C.; Varone, A. An innovative industrial process for forging 70323 Al alloy. *Mater. Sci. Forum* **2018**, *941*, 1047–1052. [CrossRef]

24. Gourdet, S.; Montheillet, F. Effects of dynamic grain boundary migration during the hot compression of high stacking fault energy metals. *Acta Mater.* **2002**, *50*, 2801–2812. [CrossRef]

25. McQueen, H.J. Development of dynamic recrystallization theory. *Mater. Sci. Eng. A* **2004**, *387–389*, 203–208. [CrossRef]

26. Humphreys, F.J.; Hatherly, M. *Recrystallization and Related Annealing Phenomena*, 2nd ed.; Elsevier: Amsterdam, The Netherlands, 2004.

27. Di Schino, A.; Di Nunzio, P.E. Metallurgical aspects related to contact fatigue phenomena in steels for back up rolling. *Acta Metall. Slovaca* **2017**, *23*, 62–71. [CrossRef]

28. Jackson, P.J. The role of cross-slip in the plastic deformation of crystals. *Mater. Sci. Eng.* **1983**, *57*, 39–47. [CrossRef]

29. Saada, G. Cross-slip and work hardening of fcc crystals. *Mater. Sci. Eng. A* **1991**, *137*, 177–183. [CrossRef]

30. Poirier, J.P. On the symmetrical role of cross-slip of screw dislocations and climb of edge dislocations as recovery process controlling high-temperature creep. *Rev. Phys. Appl.* **1976**, *11*, 731. [CrossRef]

31. Hussein, A.M.; Rao, S.I.; Uchic, M.D.; Dimiduk, D.M.; El-Awady, J.A. Microstructurally based cross-slip mechanisms and their effects on dislocation microstructure evolution in fcc crystals. *Acta Mater.* **2015**, *85*, 180–190. [CrossRef]

32. Jarzebska, A.; Bogucki, R.; Bieda, M. Influence of degree of deformation and aging time on mechanical properties and microstructure of aluminium alloy with zinc. *Arch. Metal. Mat.* **2015**, *60*, 215–221.

33. De Hass, M.; De Hosson, T.H. Grain Boundary segregation and precipitation in aluminium alloys. *Scripta Mat.* **2001**, *44*, 281–286. [CrossRef]

34. Nes, E.; Ryum, M.; Huneri, O. On the Zener drag. *Acta Metal.* **1985**, *33*, 11–22. [CrossRef]

35. Astarita, A.; Testani, C.; Scherillo, F.; Squillace, A.; Carrino, L. Beta Forging of a Ti6Al4V Component for Aeronautic Applications: Microstructure Evolution. *Metall. Microstruct. Analy.* **2014**, *3*, 460–467. [CrossRef]

Publisher's Note: MDPI stays neutral with regard to jurisdictional claims in published maps and institutional affiliations.

Article

Ti Interlayer Mediated Uniform NiGe Formation under Low-Temperature Microwave Annealing

Jun Yang [1,2], Yunxia Ping [1,*], Wei Liu [1,2], Wenjie Yu [2], Zhongying Xue [2], Xing Wei [2], Aimin Wu [2] and Bo Zhang [2,*]

[1]   School of Mathematics, Physics and Statistics, Shanghai University of Engineering Science, Shanghai 201600, China; junyang@sues.edu.cn (J.Y.); weiliusues@163.com (W.L.)

[2]   State Key Laboratory of Functional Material for Informatics, Shanghai Institute of Microsystem and Information Technology, CAS, Shanghai 200050, China; casan@mail.sim.ac.cn (W.Y.); simsnow@mail.sim.ac.cn (Z.X.); xwei@mail.sim.ac.cn (X.W.); wuaimin@mail.sim.ac.cn (A.W.)

*   Correspondence: xyping@sues.edu.cn (Y.P.); bozhang@mail.sim.ac.cn (B.Z.)

Abstract: The reactions between nickel and germanium are investigated by the incorporation of a titanium interlayer on germanium (100) substrate. Under microwave annealing (MWA), the nickel germanide layers are formed from 150 °C to 350 °C for 360 s in ambient nitrogen atmosphere. It is found that the best quality nickel germanide is achieved by microwave annealing at 350 °C. The titanium interlayer becomes a titanium cap layer after annealing. Increasing the diffusion of Ni by MWA and decreasing the diffusion of Ni by Ti are ascribed to induce the uniform formation of nickel germanide layer at low MWA temperature.

Keywords: microwave annealing; nickel germanide; titanium interlayer

check for
updates

Citation: Yang, J.; Ping, Y.; Liu, W.; Yu, W.; Xue, Z.; Wei, X.; Wu, A.; Zhang, B. Ti Interlayer Mediated Uniform NiGe Formation under Low-Temperature Microwave Annealing. *Metals* **2021**, *11*, 488. https://doi.org/10.3390/met 11030488

Academic Editor: Claudio Testani

Received: 24 December 2020
Accepted: 11 March 2021
Published: 15 March 2021

Publisher's Note: MDPI stays neutral with regard to jurisdictional claims in published maps and institutional affiliations.

1. Introduction

Germanium (Ge) has attracted a great deal of contemporary interest as a channel material in high-performance transistors, due to its higher hole and electron mobility than Si [1–3]. Due to nickel germanide (Ni_xGe_y) having the advantages of low resistivity, low formation temperature, and the feasibility of self-aligned germination process [4,5], it has been selected as the most promising contact material for Ge-based Metal-Oxide-Semiconductor Field-Effect Transistor (MOSFET) devices. However, one major drawback of Ni_xGe_y is its rough interface due to polycrystalline grains and agglomeration at high annealing temperatures (500–550 °C). It was reported that the thermal stability of Ni_xGe_y on bulk Ge could be improved by pre-germanidation implantation [6], prior-germanidation fluorine implantation into Ge substrate [7], dopant segregation [8], or introducing other elements, such as titanium (Ti) [9], platinum (Pt) [10], tungsten (W) [11], tantalum (Ta) [11,12], cobalt (Co) [13], ytterbium (Yb) [14], and so on.

On the other hand, various annealing methods have been used to prepare Ni_xGe_y films, such as rapid thermal annealing (RTA) [15] and laser annealing [16]. However, these methods have some limitations. Halogen lamp rapid thermal processing technology has problems such as high thermal budget, uneven heating, and a large residual defect density. For the laser annealing method, the spot size is small which will take a great deal of time to cover the entire surface area of a sample. Microwave heating has increasingly attracted attention in several industrial applications. For example, previous research has shown that low-temperature microwave annealing may be an alternative to other rapid thermal processing methods in silicon processing [17–19]. Oghbaei et al. reported that microwave annealing (MWA) reduces energy consumption, processing time, and annealing temperature [20]. Hu et al. reported that the formation of low-resistivity nickel germanosilide (NiSiGe) film via microwave heating occurs at temperatures about 100 °C lower than using RTA [21]. For the Ni/Ge systems, there are few studies which focus on the Ni–Ge solid-state reaction under MWA. Hsu et al. studied the effects of a Pt interlayer

49

on the structural and electrical properties of Ni_xGe_y through MWA and found the Pt interlayers played a role during the alloy formation by preventing rapid Ni diffusion [22]. However, the formation of a Ni_xGe_y layer in the appearance of other metals under MWA has not been fully investigated.

In this study, we investigated the formation of Ni_xGe_y layers by Ti incorporation on a Ge (100) substrate under MWA. It was found that the formation of uniform nickel germanide layer could be achieved under MWA at 350 °C. Most of the Ti atoms moved to the surface of the NiGe layer after annealing. The mechanism analysis of MWA and the effects of Ti mediation are discussed in detail.

2. Experimental

Pure Ge(001) substrates were used in this study. The Radio Corporation of America (RCA) standard clean 1 (SC1), consisting of $NH_4OH:H_2O_2:H_2O$ at a 1/2/20 ratio and temperature of 60 °C for 10 min was used to remove the organics, certain metals, and particles. The RCA standard clean 2 (SC2), consisting of $HCl:H_2O_2:H_2O$ at 1/1/20 ratio and temperature of 60 °C for 10 min is used to remove metallic contamination. In order to remove germanium oxides on the surface, the Ge substrates were cleaned using a dilute fluoric acid (1% HF) for 60 s. Then, a 1 nm Ti intermediate layer and a 10 nm Ni layer were successively deposited on the Ge (100) substrate by using electron beam evaporation. After the deposition, these wafers were sliced into pieces of 2.0 cm × 2.0 cm. Then, the samples were annealed in an AXOM-200 (Manufactured by DSG Technologies) MWA chamber (5.8 GHz) at 150 °C, 200 °C, 250 °C, 300 °C, and 350 °C for 360 s in an ambient nitrogen atmosphere. The temperature image after calibration and the error of each temperature are shown in Appendix A (Figure A1 and Table A1). The MWA chamber is designed for multi-wafer processing; the vertically stacked wafers are supported by three quartz rods inside a quartz chamber. The card slot on the quartz column can satisfy the annealing of several wafers together. The three adjustable quartz columns make the equipment suitable for 4, 6, and 8 inch wafer annealing. All samples were located inside the middle of the chamber where the electromagnetic field was uniform. An infrared pyrometer located at the bottom of the chamber was used to directly monitor the sample temperature. Different temperatures could be achieved by adjusting the input power to generate continuous microwave output from 10% to 100% of the maximum rated microwave power output. To avoid oxidation, the nitrogen flow was maintained until the annealing process was completed.

The as-prepared Ni_xGe_y samples were characterized by four-point probe (FPP) measurements for sheet resistance (R_{sh}), Raman spectroscopy (Raman) for phase formation identification, transmission electron microscopy (TEM) for morphology and microstructure observations, and energy dispersive spectrometer (EDS) for elemental mapping and line scans in Ni_xGe_y films.

3. Results and Discussion

3.1. Characterization of the Ni_xGe_y Layers

The influence of MWA on the nickel germanide formed on a Ge substrate was studied by measuring sheet resistance. The variation of sheet resistance (R_{sh}) with germanidation temperature is shown in Figure 1a. At 150 °C, the sheet resistance value is very large after etching, which indicates that Ni has not reacted with Ge. This result could be also confirmed by the Raman analysis, as shown in Figure 2. The sheet resistance value decreased gradually after annealed at 200 °C, indicating the Ni-rich phase (Ni_2Ge or Ni_5Ge_3) with higher resistance transition to the Ni mono-nickel germanide phase (NiGe) [23–25]. In addition, in order to calculate the morphology and roughness of the Ni_xGe_y layers, AFM roughness analysis was carried out, as shown in Figure 1b,c. According to the AFM measurements, it was found that the films were smooth, and the root mean square surface roughness increased slightly from 1.29 nm to 1.36 nm.

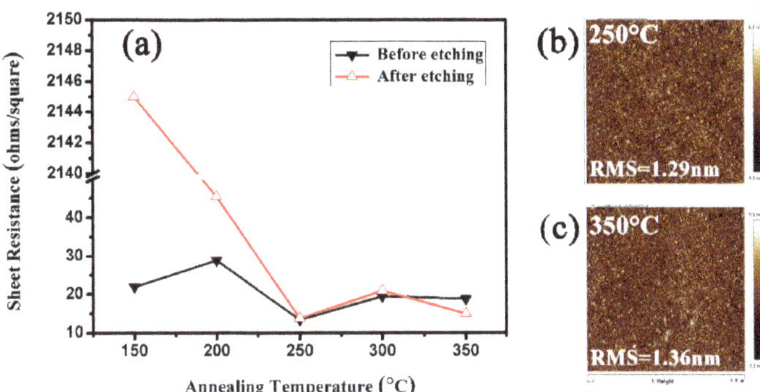

Figure 1. Sheet resistance of Ni/Ti/Ge samples annealed at various temperatures. (**a**) AFM surface morphologies with the scanning area of $5 \times 5\ \mu m^2$ of (**b**) 250 °C, (**c**) 350 °C annealed sample.

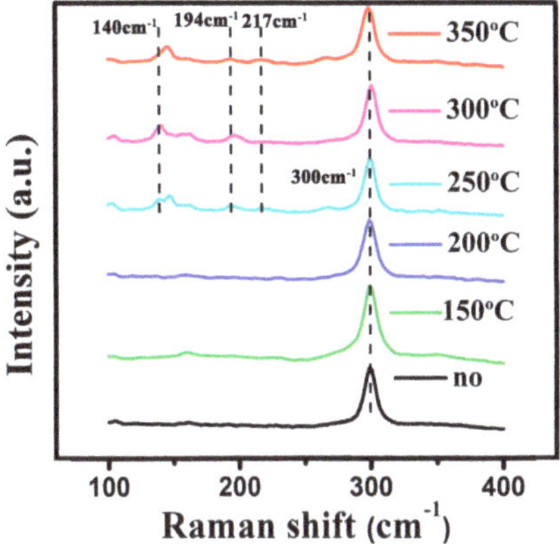

Figure 2. Raman spectra of Ni_xGe_y films formed at various temperatures.

Figure 2 shows the Raman spectra of the as-deposited and annealed samples at temperatures ranging from 150 to 350 °C. For the as-deposited and 150/200 °C annealed samples, no Ni–Ge bond vibrational Raman signal is observed and only the characteristic peak of Ge–Ge bond vibration is shown at 300 cm^{-1}, which indicates that no Ni_xGe_y phase is formed at these temperatures. After annealing from 250–350 °C, three main Raman peaks are found to be approximately located at 140 cm^{-1}, 194 cm^{-1} and 217 cm^{-1}, which could be attributed to the NiGe phase [26,27].

In order to further study the structures of germanide layer, the images of high-angle annular dark field scanning TEM (HAADF-STEM) and EDS mappings of the stacking structure are shown in Figure 3, respectively. At 250 °C, there are some grains in the germanide layer, which induced non-uniform Ni element distribution. In addition, it was found that there was some un-reacted Ni on the surface of the germanide layer. It shows that after annealing at 250 °C, Ni had not completely reacted with Ge. After annealing at a relative higher temperature 350 °C, we noticed that there was a clear interface between the Ge substrate and the germanide layer. The NiGe/Ge interface was ideal, and NiGe layer was very smooth without grains. Moreover, as shown in the EDS mapping, for the region which the Ni and Ge mixed, a significant Ni distribution indicates the formation of a nickel germanide layer. It can be seen from the image that the Ti atoms have diffused into the sample surface, which changed from the interlayer to the surface coating of the sample. Compared with the sample annealing at 250 °C, the titanium layer on the surface is smoother, which indicates that the film has good uniformity after annealing at 350 °C.

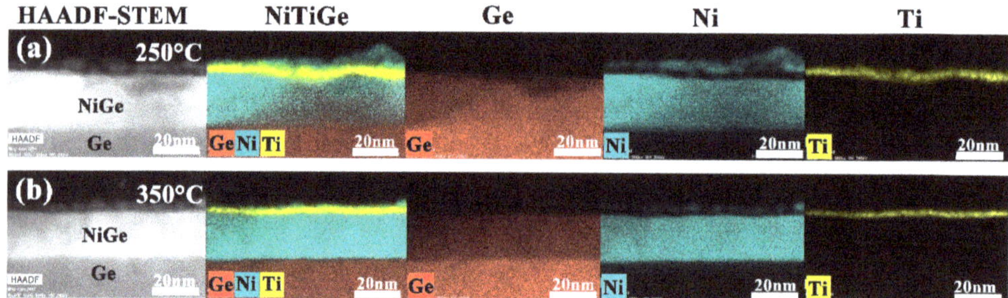

Figure 3. HAADF-STEM and EDS mapping images of the nickel germanide film layer at various temperatures: (**a**) 250 °C, (**b**) 350 °C.

Figure 4 shows the high-resolution TEM (HRTEM) images of Ni/Ti/Ge samples after MWA at 350 °C. It was found that the NiGe layer was continuous and uniform. The enlarged view shows the NiGe lattice interface image had a good polycrystalline structure and its thickness was about 20 nm. Relative uniformity of the NiGe film and a distinct interface between the NiGe and Ge could be observed. Combining the results of Figures 3b and 5b, we believe that a uniform NiGe layer can be obtained under MWA at 350 °C.

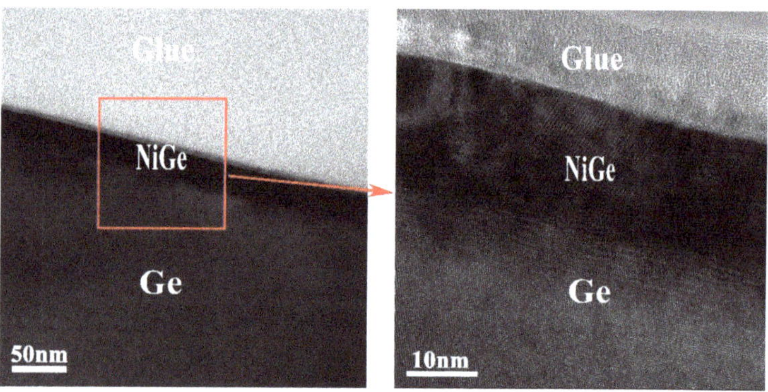

Figure 4. HRTEM images of NiGe films annealed at 350 °C.

For Ni_xGe_y films formed with MWA at 250 °C and 350 °C, the EDS measurement depth profiles for different elements are displayed in Figure 5a,b, respectively. We found that Ti atoms were mainly located in the surface area of the Ni_xGe_y film when Ni reacted with Ge to form germanide, which is consistent with our previous reports [28]. The EDS line profiles suggest that in the region where the Ni and Ge mixed, the 20 nm nickel germanide layer was formed, which is consistent with the HRTEM results, as shown in Figure 4. For the MWA sample at 250 °C, a nickel peak appeared on the surface of the sample and the signal distribution of Ni and Ge in the middle part was uneven, indicated that some Ni-rich phases were formed (see Figure 5a). However, for the 350 °C MWA sample, as shown in Figure 5b, the Ni and Ge intensity was uniformly distributed, which confirmed that the formation of the mono-germanide phase (NiGe).

3.2. Mechanism Analysis of MWA

For NiGe formed on Ge, we suspect that dielectric loss and conductivity loss are two possible properties responsible for the microwave losses during microwave annealing [21]. For the dielectric loss, it is caused not only by the lattice vibrational modes, but also by impurities, second phases, pores, lattice defects, grain boundaries, and grain morphology [29]. The point defects generated in Ni–Ge interactions and the grain boundaries of polycrystalline NiGe grains could affect the dielectric loss under MWA. For the conductivity loss, the layers stacked in our samples Ni/Ti/Ge are expected to be heated differently, because the loss factor of different materials determines its absorption of the propagating electric field. According to Ampere's and Faraday's laws [30], the effectiveness of electromagnetic field penetration depends on a parameter referred to as the skin depth. With the increase in conductivity, skin depth decreases, and heating can occur in the skin depth layer [31]. Due to the conductivity of Ni (1.46×10^7 s/m) being much greater than that of Ge (2 s/m) [32], the temperature of the Ni layer is higher than the Ge substrate under microwave heating. The high temperature of the Ni layer will also lead to the enhancement of the Ni diffusion rate.

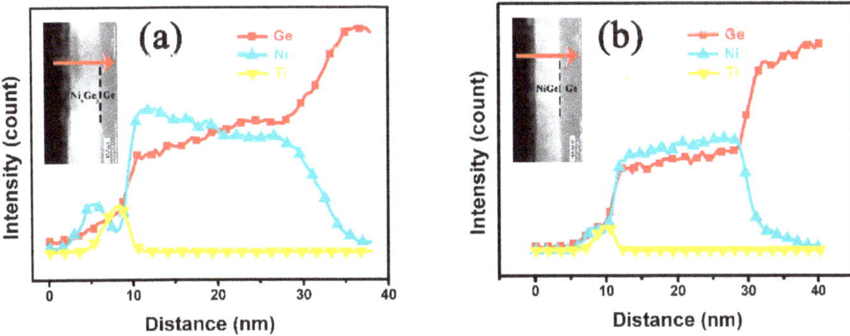

Figure 5. EDS profiles of Ni, Ge, and Ti from the Ni$_x$Ge$_y$ layer annealed at (**a**) 250 °C, (**b**) 350 °C. The inset shows the scan direction of the EDS measurements.

3.3. The Benefits of MWA

Previous research [21] has shown that the local temperature at the substrate front surface where silicidation occurs is higher for the MWA than RTA samples. A local temperature difference at the reaction interface is a reason for the dissimilarity between MWA and RTA. For the RTA process, it is well known that the Ni–Ge reaction starts at 250 °C and a Ti thin film reacts on a Ge substrate above 400 °C [33]. However, under MWA, we found that the mono-germanide phase can be formed at 350 °C in the Ni/Ti/Ge system. The contribution of microwave energy to overcome the activation energy (Ea) is an important factor for NiGe formation, because Ea could be overcome by the combination of substrate temperature and direct coupling of microwave energy into the lattice [34]. Thus, the mono-NiGe phase formed at lower temperatures, similar to the formation of low-resistivity NiSiGe film via MWA which occurs at temperatures about 100 °C lower than using RTA [21].

3.4. The Effect of Ti Mediation under MWA

Zhu et al. reported that the Ti interlayer is expelled to the surface and forms a ternary Ni$_{1-x}$Ti$_x$Ge phase under RTA 450 °C [9]. Under MWA, the 1 nm Ti interlayer is also totally moved to the surface. However, Ti does not react with Ge in the Ni/Ti/Ge system under MWA at 350 °C, corresponding to the findings that a Ti layer could only react with Ge substrates above 400 °C [33]. The Ti interlayer becomes a perfect cap-layer on the top of NiGe, which could further "protect" the contact property of NiGe. Moreover, it is reported that the Pt interlayer plays a role in preventing the rapid diffusion of Ni and forms PtGe(Ni) and NiGe(Pt) in Ni/Pt/Ge system [22] under the MWA process. The Ti interlayer, like the Pt interlayers, may act as a diffusion barrier to change the Gibbs energy of the NiGe grains and reduce the NiGe growth rate. Similar phenomena have also been found in the cases of Ni/Ti/Si [34], Ni/Ti/SiGe [28] and Ni/Ti/Ge [9] systems.

Based on the above discussion, we propose that two effects, increasing the diffusion of Ni-Ge by MWA and decreasing the diffusion of Ni-Ge by Ti, both occur and reach a balance at some MWA temperatures. However, the detailed mechanisms of Ti on the formation of NiGe under MWA are still not fully understood and require further investigations.

4. Conclusions

In summary, we have investigated the reactions between Ni and Ge in the appearance of a Ti interlayer under MWA processed. The variation of sheet resistance, evolution of surface, and interfacial morphology with different germanidation temperature were investigated systematically. It was demonstrated that by mediation of the Ti interlayer, a flat and uniform NiGe layer was formed at 350 °C. The Ti atoms were found to segregate at the surface of NiGe after annealing. It as shown that MWA is a viable alternative method for the formation of NiGe films at low temperatures. Achieved results in this work could be a potential precursor for S/D contact technology in state-of-the-art Ge-based devices.

Author Contributions: Conceptualization, J.Y., Y.P. and B.Z.; data curation, J.Y.; funding acquisition, Y.P. and B.Z.; investigation, J.Y. and W.L.; supervision and project administration, Y.P., W.Y., Z.X., X.W., A.W. and B.Z.; writing—original draft preparation, J.Y.; writing—review and editing, Y.P. and B.Z.; All authors have read and agreed to the published version of the manuscript.

Funding: This research was funded by the National Natural Science Foundation of China, grant numbers 61604094.

Institutional Review Board Statement: Not applicable.

Informed Consent Statement: Not applicable.

Data Availability Statement: The data presented in this study are available on request from the corresponding author.

Conflicts of Interest: The authors declare no conflict of interest.

Appendix A

In our experiment, MWA equipment used an infrared pyrometer to measure the temperature of the sample. However, in the non-contact infrared temperature measurement, the measured temperature deviated from the actual temperature, so it was necessary to calibrate the temperature. The temperature image after calibration is shown in Figure A1. In addition, due to the limitation of the infrared pyrometer (especially for the low temperature), we provide an error for each temperature, as shown in Table A1.

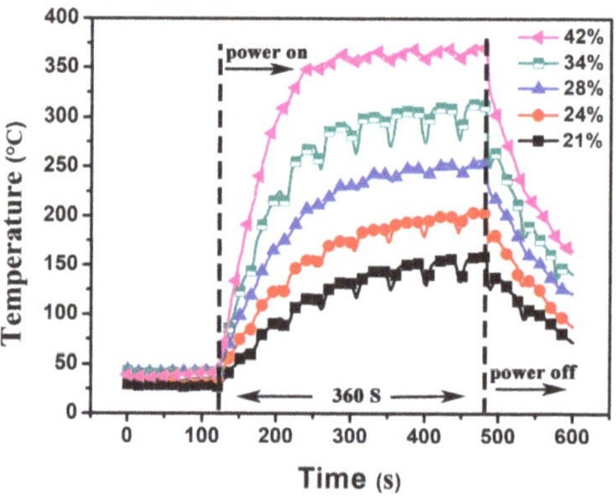

Figure A1. Annealing temperature curve of the actual sample after calibration.

Table A1. Temperature deviation in microwave annealing.

Microwave Annealing Temperature (°C)	Minimum Temperature (°C)	Maximum Temperature (°C)
150	130	160
200	173	202
250	233	257
300	277	314
350	342	366

References

1. Chi On, C.; Hyoungsub, K.; Chi, D.; Triplett, B.B.; McIntyre, P.C.; Saraswat, K.C. A sub-400/spl deg/C germanium MOSFET technology with high-/spl kappa/dielectric and metal gate, Digest. In Proceedings of the International Electron Devices Meeting, San Francisco, CA, USA, 8–11 December 2002; IEEE: New York, NY, USA, 2002; pp. 437–440. [CrossRef]
2. Huiling, S.; Okorn-Schmidt, H.; Chan, K.K.; Copel, M.; Ott, J.A.; Kozlowski, P.M.; Steen, S.E.; Cordes, S.A.; Wong, H.P.; Jones, E.C.; et al. High mobility p-channel germanium MOSFETs with a thin Ge oxynitride gate dielectric, Digest. In Proceedings of the International Electron Devices Meeting, San Francisco, CA, USA, 8–11 December 2002; IEEE: New York, NY, USA, 2002; pp. 441–444. [CrossRef]
3. Zhu, S.; Li, R.; Lee, S.J.; Li, M.F.; Du, A.; Singh, J.; Zhu, C.; Chin, A.; Kwong, D.L. Germanium pMOSFETs with Schottky-barrier germanide S/D, high-/spl kappa/gate dielectric and metal gate. *IEEE Electron Device Lett.* **2005**, *26*, 81–83. [CrossRef]
4. Hsu, S.-L.; Chien, C.-H.; Yang, M.-J.; Huang, R.-H.; Leu, C.-C.; Shen, S.-W.; Yang, T.-H. Study of thermal stability of nickel monogermanide on single- and polycrystalline germanium substrates. *Appl. Phys. Lett.* **2005**, *86*, 251906. [CrossRef]
5. Zhang, Q.; Wu, N.; Osipowicz, T.; Bera, L.K.; Zhu, C. Formation and Thermal Stability of Nickel Germanide on Germanium Substrate. *Jpn. J. Appl. Phys.* **2005**, *44*, L1389–L1391. [CrossRef]
6. Liu, Y.; Xu, J.; Liu, J.; Wang, G.; Luo, X.; Zhang, D.; Mao, S.; Li, Y.; Li, J.; Zhao, C.; et al. Role of Carbon Pre-Germanidation Implantation on Enhancing the Thermal Stability of NiGe Films Below 10 nm Thickness. *ECS J. Solid State Sci. Technol.* **2020**, *9*, 054006. [CrossRef]
7. Duan, N.; Wang, G.; Xu, J.; Mao, S.; Luo, X.; Zhang, D.; Wang, W.; Chen, D.; Li, J.; Liu, S.; et al. Enhancing the thermal stability of NiGe by prior-germanidation fluorine implantation into Ge substrate. *Jpn. J. Appl. Phys.* **2018**, *57*, 07MA03. [CrossRef]
8. Chen, Z.; Yuan, S.; Li, J.; Zhang, R. Thermal Stability Enhancement of NiGe Metal Source/Drain and Ge pMOSFETs by Dopant Segregation. *IEEE Trans. Electron Devices* **2019**, *66*, 5284–5288. [CrossRef]
9. Zhu, S.; Yu, M.B.; Lo, G.Q.; Kwong, D.L. Enhanced thermal stability of nickel germanide on thin epitaxial germanium by adding an ultrathin titanium layer. *Appl. Phys. Lett.* **2007**, *91*, 051905. [CrossRef]
10. Nakatsuka, O.; Suzuki, A.; Sakai, A.; Ogawa, M.; Zaima, S. Impact of Pt Incorporation on Thermal Stability of NiGe Layers on Ge(001) Substrates. In Proceedings of the 2007 International Workshop on Junction Technology, Kyoto, Japan, 8–9 June 2007; IEEE: New York, NY, USA, 2007; pp. 87–88. [CrossRef]
11. Jablonka, L.; Kubart, T.; Gustavsson, F.; Descoins, M.; Mangelinck, D.; Zhang, S.L.; Zhang, Z. Improving the morphological stability of nickel germanide by tantalum and tungsten additions. *Appl. Phys. Lett.* **2018**, *112*, 103102. [CrossRef]
12. Lee, J.W.; Kim, H.K.; Bae, J.H.; Park, M.H.; Kim, H.; Ryu, J.; Yang, C.W. Enhanced morphological and thermal stabilities of nickel germanide with an ultrathin tantalum layer studied by ex situ and in situ transmission electron microscopy. *Microsc. Microanal.* **2013**, *19*, 114–118. [CrossRef]
13. Shin, G.-H.; Kim, J.; Li, M.; Lee, J.; Lee, G.-W.; Oh, J.; Lee, H.-D. A Study on Thermal Stability Improvement in Ni Germanide/p-Ge using Co interlayer for Ge MOSFETs. *J. Semicond. Technol. Sci.* **2017**, *17*, 277–282. [CrossRef]
14. Zhang, Y.-Y.; Oh, J.; Li, S.-G.; Jung, S.-Y.; Park, K.-Y.; Shin, H.-S.; Lee, G.-W.; Wang, J.-S.; Majhi, P.; Tseng, H.-H.; et al. Ni Germanide Utilizing Ytterbium Interlayer for High-Performance Ge MOSFETs. *Electrochem. Solid-State Lett.* **2009**, *12*, H18–H20. [CrossRef]
15. Lee, K.Y.; Liew, S.L.; Chua, S.J.; Chi, D.Z.; Sun, H.P.; Pan, X.Q. Formation and Morphology Evolution of Nickel Germanides on Ge (100) under Rapid Thermal Annealing. *MRS Proceed.* **2004**, *810*, C2.4. [CrossRef]
16. Lim, P.S.Y.; Chi, D.Z.; Lim, P.C.; Wang, X.C.; Chan, T.K.; Osipowicz, T.; Yeo, Y.-C. Formation of epitaxial metastable NiGe$_2$ thin film on Ge(100) by pulsed excimer laser anneal. *Appl. Phys. Lett.* **2010**, *97*, 182104. [CrossRef]
17. Kappe, C.O. Controlled microwave heating in modern organic synthesis. *Angew. Chem. Int. Ed.* **2004**, *43*, 6250–6285. [CrossRef]
18. Lee, Y.J.; Hsueh, F.K.; Current, M.I.; Wu, C.Y.; Chao, T.S. Susceptor coupling for the uniformity and dopant activation efficiency in implanted Si under fixed-frequency microwave anneal. *IEEE Electron Device Lett.* **2012**, *33*, 248–250. [CrossRef]
19. Alford, T.L.; Thompson, D.C.; Mayer, J.W.; Theodore, N.D. Dopant activation in ion implanted silicon by microwave annealing. *J. Appl. Phys.* **2009**, *106*. [CrossRef]
20. Oghbaei, M.; Mirzaee, O. Microwave versus conventional sintering: A review of fundamentals, advantages and applications. *J. Alloys Compd.* **2010**, *494*, 175–189. [CrossRef]

21. Hu, C.; Xu, P.; Fu, C.; Zhu, Z.; Gao, X.; Jamshidi, A.; Noroozi, M.; Radamson, H.; Wu, D.; Zhang, S.-L. Characterization of Ni(Si,Ge) films on epitaxial SiGe(100) formed by microwave annealing. *Appl. Phys. Lett.* **2012**, *101*, 092101. [CrossRef]
22. Hsu, C.-C.; Lin, K.-L.; Chi, W.-C.; Chou, C.-H.; Luo, G.-L.; Lee, Y.-J.; Chien, C.-H. Phase-separation phenomenon of NiGePt alloy on n-Ge by microwave annealing. *J. Alloys Compd.* **2018**, *743*, 262–267. [CrossRef]
23. De Schutter, B.; Van Stiphout, K.; Santos, N.M.; Bladt, E.; Jordan-Sweet, J.; Bals, S.; Lavoie, C.; Comrie, C.M.; Vantomme, A.; Detavernier, C. Phase formation and texture of thin nickel germanides on Ge(001) and Ge(111). *J. Appl. Phys.* **2016**, *119*, 135305. [CrossRef]
24. Jablonka, L.; Kubart, T.; Primetzhofer, D.; Abedin, A.; Hellström, P.-E.; Östling, M.; Jordan-Sweet, J.; Lavoie, C.; Zhang, S.-L.; Zhang, Z. Formation of nickel germanides from Ni layers with thickness below 10 nm. *J. Vac. Sci. Technol. B* **2017**, *35*, 020602. [CrossRef]
25. Begeza, V.; Mehner, E.; Stocker, H.; Xie, Y.; Garcia, A.; Hubner, R.; Erb, D.; Zhou, S.; Rebohle, L. Formation of Thin NiGe Films by Magnetron Sputtering and Flash Lamp Annealing. *Nanomaterials* **2020**, *10*, 648. [CrossRef]
26. Guo, Y.; An, X.; Huang, R.; Fan, C.; Zhang, X. Tuning of the Schottky barrier height in NiGe/n-Ge using ion-implantation after germanidation technique. *Appl. Phys. Lett.* **2010**, *96*, 143502. [CrossRef]
27. Uddin, W.; Saleem Pasha, M.; Dhyani, V.; Maity, S.; Das, S. Temperature dependent current transport behavior of improved low noise NiGe schottky diodes for low leakage Ge-MOSFET. *Semicond. Sci. Technol.* **2019**, *34*, 035026. [CrossRef]
28. Ping, Y.; Hou, C.; Zhang, C.; Yu, W.; Xue, Z.; Wei, X.; Peng, W.; Di, Z.; Zhang, M.; Zhang, B. Ti mediated highly oriented growth of uniform and smooth Ni(Si$_{0.8}$Ge$_{0.2}$) layer for advanced contact metallization. *J. Alloys Compd.* **2017**, *693*, 527–533. [CrossRef]
29. Tamura, H. Microwave dielectric losses caused by lattice defects. *J. Eur. Ceram. Soc.* **2006**, *26*, 1775. [CrossRef]
30. Shen, L.C.; Kong, J.A. *Applied Electromagnetism*; PWS Publishing: Boston, MA, USA, 1995; p. 552.
31. Wang, T.; Dai, Y.; Dai, Q.; Ng RM, Y.; Chan, W.T.; Lee, P.; Chan, M. Microwave plasma anneal to fabricate silicides and restrain the formation of unstable phases. In Proceedings of the 2007 IEEE Conference on Electron Devices and Solid-State Circuits, Tainan, Taiwan, 20–22 December 2007; IEEE: New York, NY, USA, 2007; pp. 609–612. [CrossRef]
32. Moulder, J.C.; Uzal, E.; Rose, J.H. Thickness and conductivity of metallic layers from eddy current measurements. *Rev. Sci. Instrum.* **1992**, *63*, 3455–3465. [CrossRef]
33. Gaudet, S.; Detavernier, C.; Kellock, A.J.; Desjardins, P.; Lavoie, C. Thin film reaction of transition metals with germanium. *J. Vac. Sci. Technol.* **2006**, *24*, 474–485. [CrossRef]
34. Nakatsuka, O.; Okubo, K.; Tsuchiya, Y.; Sakai, A.; Zaima, S.; Yasuda, Y. Low-Temperature Formation of Epitaxial NiSi$_2$ Layers with Solid-Phase Reaction in Ni/Ti/Si(001) Systems. *Jpn. J. Appl. Phys.* **2005**, *44*, 2945–2947. [CrossRef]

Article

Casting Microstructure Inspection Using Computer Vision: Dendrite Spacing in Aluminum Alloys

Filip Nikolić [1,2,3], Ivan Štajduhar [4,*] and Marko Čanađija [1,*]

1   Department of Engineering Mechanics, Faculty of Engineering, University of Rijeka, 51000 Rijeka, Croatia; fnikolic@riteh.hr
2   Research and Development Department, CIMOS d.d. Automotive Industry, 6000 Koper, Slovenia
3   CAE Department, Elaphe Propulsion Technologies Ltd., 1000 Ljubljana, Slovenia
4   Department of Computer Engineering, Faculty of Engineering, University of Rijeka, 51000 Rijeka, Croatia
*   Correspondence: istajduh@riteh.hr (I.Š.); marko.canadija@riteh.hr (M.Č.); Tel.: +385-51-651-448 (I.Š.); +385-51-651-496 (M.Č.); Fax: +385-51-651-416 (I.Š.); +385-51-651-490 (M.Č.)

Abstract: This paper investigates the determination of secondary dendrite arm spacing (SDAS) using convolutional neural networks (CNNs). The aim was to build a Deep Learning (DL) model for SDAS prediction that has industrially acceptable prediction accuracy. The model was trained on images of polished samples of high-pressure die-cast alloy EN AC 46000 AlSi9Cu3(Fe), the gravity die cast alloy EN AC 51400 AlMg5(Si) and the alloy cast as ingots EN AC 42000 AlSi7Mg. Color images were converted to grayscale to reduce the number of training parameters. It is shown that a relatively simple CNN structure can predict various SDAS values with very high accuracy, with a R^2 value of 91.5%. Additionally, the performance of the model is tested with materials not used during training; gravity die-cast EN AC 42200 AlSi7Mg0.6 alloy and EN AC 43400 AlSi10Mg(Fe) and EN AC 47100 Si12Cu1(Fe) high-pressure die-cast alloys. In this task, CNN performed slightly worse, but still within industrially acceptable standards. Consequently, CNN models can be used to determine SDAS values with industrially acceptable predictive accuracy.

Keywords: secondary dendrite arm spacing; convolutional neural network; casting microstructure inspection; deep learning; aluminum alloys

Citation: Nikolić, F.; Štajduhar, I.; Čanađija, M. Casting Microstructure Inspection Using Computer Vision: Dendrite Spacing in Aluminum Alloys. *Metals* **2021**, *11*, 756. https://doi.org/10.3390/met11050756

Academic Editor: Claudio Testani

Received: 13 April 2021
Accepted: 29 April 2021
Published: 4 May 2021

Publisher's Note: MDPI stays neutral with regard to jurisdictional claims in published maps and institutional affiliations.

1. Introduction

It is well known that the size of dendrites and the secondary dendrite arm spacing (SDAS) strongly depend on the solidification rate of a given material [1,2]. In addition, the chemical composition of the alloys has an additional influence on this structural characteristic [3]. Moreover, some authors have shown a relationship between mechanical properties and SDAS [1,4–8]. Properties of fracture mechanics will also depend on chemical composition, casting defects such as porosity and oxide films [8], and the size and shape of Si or Fe-rich brittle phases [9]. Most authors show a relationship between SDAS and ultimate tensile strengths (UTS) and elongation (E), while many authors indicate that SDAS has no significant effect on yield strength (YS). Moreover, another study indicates that the hardness of the material depends on SDAS, but cannot be described well enough using only this relationship [10]. Consequently, it is reasonable to assume that some material properties can be determined directly from the value of SDAS. Thus, it could be useful to know the SDAS value of the material. In this regard, an automatic method for determining SDAS could be a significant advantage.

The scope of artificial intelligence (AI) is more significant in disciplines such as computer science or electrical engineering than in materials science. However, in the last three decades, many applications can be found in materials science as well. In general, neural networks, as a core algorithm of AI, have been applied in materials science as early as 1998 [11]. Singh et al. estimated the YS and UTS as a function of each of the 108 variables for the steel rolling

57

process. An artificial neural network (ANN) with only two hidden units was used. In another early paper, Hancheng et al. [12] developed a model to predict tensile strength based on compositions and microstructure by using adaptive neuro-fuzzy inference method. Agrawal et al. [13] determined the fatigue strength from steel processing parameters with R^2 score of over 97%. Liu et al. [14] determined the Young's modulus, YS, and magnetostrictive strain using machine learning techniques (ML). The ML approach reduces the computation time by an average of 81.62% compared to standard optimization methods. Yang et al. [15] used ANN with one layer and 8 neurons to predict UTS and E from the heat treatment parameters. Santos et al. [16] predicted UTS of an automotive iron casting using ANN and K-Nearest Neighbor method, which outperformed several other methods, with the highest accuracy reaching 85%. Liao et al. [17] constructed ML algorithms to predict macroshrinkage of aluminum alloys based on their experimental dataset. Additionally, the interested reader is referred to reference [18], which shows several applications of DL in the field of materials science. Furthermore, in [18] and references therein, determination of material properties using DL could be found as well. Herriott and Spear [19] investigate the ability of ML and DL models to predict microstructure-sensitive mechanical properties in additive manufacturing of metals (MAM). ML and DL methods accelerated the prediction of microstructure and mechanical response for MAM.

Given that in recent decades our ability to generate data has far surpassed our ability to make sense of it in virtually all scientific domains [13], the development of DL methods could be of particular benefit in materials science. One process that involves a lot of input data is quality control. Quality control is a fundamental part of many manufacturing processes, especially casting or welding. Unfortunately, manual quality control procedures are often time consuming and error prone [20]. Quality control, SDAS in particular, could also be performed by Light Optical Microscopy (LOM) of polished samples, but SDAS evaluation still relies on manual measurements. Although there exists research tackling the problem of SDAS prediction using ANNs [21], SDAS is not predicted directly from the microstructure image. Instead, the authors predicted SDAS based on processing parameters: pouring temperature, insulation on the riser and chill specific heat, while the dataset was based on numerical simulation results. To the best of the authors' knowledge, there is currently no literature that determines SDAS directly from the microstructure image. Following the literature review, this research hypothesizes that SDAS could be determined directly from the microstructure image data using DL methods.

2. Related Work

The algorithms available for implementing ML can be broadly categorized into the following types: shallow learning (e.g., vector machine (SVM), decision tree (DT) and ANN) and DL (e.g., CNN, recurrent neural network (RNN), deep belief network (DBN) and deep coding network) [22]. Shallow learning algorithms cannot achieve the same accuracy on different tasks as DL, although they could reduce the computational cost. As shown in [19,23] CNN performs better than standard (2D) ML models such as Ridge, XGBoost regression and the like. A transfer learning-based approach to use convolutional deep networks in [24] is generally superior to all other reconstruction approaches (decision tree, Gaussian random field, two-point correlation, physical descriptor) for most numerically evaluated material systems. CNN has also been used for casting defects recognition in [25]. These authors used 640,000 images to train CNN and also Generative Adversarial Network (GAN) was used to generate even more data. In the present study, image data were used to predict SDAS using CNNs. It is reiterated that the task of determining SDAS is often manual and subjective.

Data used in materials science can be obtained from the following sources: Material properties from experiments and simulations, chemical reaction data, image data and published data [22]. DeCost et al. [26] applied a deep CNN segmentation model to enable novel automated microstructure segmentation applications for complex microstructures. Ferguson et al. [20] proposed a casting defect detection system in X-ray images based on the Mask Region-based CNN architecture. The proposed defect detection system outperforms the state-of-the-art performance for defect detection on GDXray Castings dataset. Chowdhury et al. [27] created an image-driven machine learning approach to classify micrographs. In references therein, classification of precipitate shapes in a two-phase microstructure, the classification of cast iron microstructures and the analysis of precipitate shapes in nickel-based superalloys could also be found. Azimi et al. [23] use a segmentation-based approach based on Fully Convolutional Neural Networks (FCNNs), which are an extension of CNNs, accompanied by a max-voting scheme to classify microstructures. Exl et al. [28] used an ML-approach to identify the importance of microstructure properties in causing magnetization reversal in ideally structured large-grained Nd2Fe14B permanent magnets. The embedded Stoner–Wohlfarth method is used as a reduced-order model to determine local switching field maps that guide the data-driven learning procedure. They used datasets consisting of 700 and 800 experiments. Yucel et al. [29] showed relationships between optical micrographs and mechanical properties (YS, UTS and E) of cold-rolled high-strength low-alloy steels (HSLA) measured in standardized tensile tests. However, the models developed in the latter work are only applicable for a constant initial condition before annealing treatment, i.e., constant composition, constant hot rolling and constant cold rolling parameters. Li et al. [24] used a transfer learning approach for microstructure reconstruction and structure-property prediction. Pokuri et al. [30] show a data-driven approach for mapping microstructure to photovoltaic performance using CNNs. The VGG-16 CNN-based architecture achieved a test accuracy of up to 96.61%. DeCost et al. [31] trained an SVM to classify microstructures into one of seven groups with accuracy greater than 80% with 5-fold cross-validation.

3. Materials and Models

3.1. Aluminum Alloy Samples

In the present study, samples for training were cut and polished from three different materials cast by three different methods: high-pressure die-cast EN AC 46000 AlSi9Cu3(Fe), gravity die-cast EN AC 51400 AlMg5(Si) and EN AC 42000 AlSi7Mg alloy cast as ingots. The chemical composition of all the alloys used conformed to DIN EN 1706 [32]. The die-cast alloys were supplied in the form of commercial ingots. The ingots were melted in the gas furnace, then degassed with argon and transferred to the holding furnace at a temperature of $700 \pm 10\ °C$. The melt was then automatically ladled into the shot sleeve and injected into the die cavity in the case of the AlSi9Cu3(Fe) alloy. Cast parts were produced using a cold chamber high pressure die casting machine (HPDC) with an intensification pressure of 800 bar. In the case of AlMg5(Si) alloy, the melt was ladled directly into the metal die and solidified under gravity conditions. Moreover, in the case of AlSi7Mg alloy, the melt was poured directly into the ingot die while exact process parameters and temperatures are not known. It should be noted that the AlMg(Si) and AlSi9Cu3(Fe) samples were cut from medium sized components between 5 and 10 kg.

Additional samples were obtained to test the prediction accuracy. This dataset consists of three materials with chemical composition according to the DIN EN 1706 [32] standard: EN AC 42200 AlSi7Mg0.6, EN AC 43400 AlSi10Mg(Fe) and EN AC 47100 AlSi12Cu1(Fe). AlSi7Mg0.6 was cast by gravity die casting technique, while AlSi10Mg(Fe) and AlSi12Cu1(Fe) alloys were cast by HPDC process. The exact process parameters and casting history for these materials are not known. Polished samples for this group of materials were cut from medium sized cast components between 5 and 10 kg.

All specimens were cut using a classic band saw and polished using standard techniques. Images were taken using an Olympus BX51 optical microscope equipped with an XC30 digital camera.

3.2. Dataset and Image Preprocessing

The dataset is created from a single polished cross section per material. The manually measured values of SDAS for AlSi9Cu3(Fe), AlSi7Mg and AlMg5(Si) alloys were $6 \cdot 10^{-3}$,

$29 \cdot 10^{-3}$ and $32 \cdot 10^{-3}$ mm, respectively. Manual SDAS measurements were performed such that the distance between the centers of two secondary dendrite arms was measured perpendicular to the primary arm, Figure 1. In other words, "Method E" from reference [33], was adopted here for manual SDAS measurement. The automated technique proposed here is based on the same SDAS measurement principle. These SDAS values were obtained at $5\times$ magnification. To obtain more SDAS values for training, additional SDAS values were derived using different magnifications on the microscope. Thus, the equation for the derived SDAS values could be obtained with the following expression:

$$SDAS = S \cdot F, \tag{1}$$

where SDAS is the physical SDAS value of the alloy, S is the scaled SDAS value used with the ML algorithm, and F is the factor of magnification. Note that the S values were used with the ML algorithm instead of the physical SDAS values. Only for magnification of $5\times$ when $F = 1$, $S = $ SDAS. The magnification factor for all magnifications used in the present research is given in Table 1. In addition, the values for the number of pixel per micrometer at different magnifications and the factor F are given. The procedure is illustrated using the example of the material AlSi7Mg in Figure 2. For this example, S values of $29 \cdot 10^{-3}$ and $55 \cdot 10^{-3}$ were obtained for training with the same polished cross-section sample with relatively constant SDAS values. The S values used for training are shown in Table 2 and Figure 3. It should be noted that S was determined using average measurement results on a polished cross section and not using Equation (1), i.e., 10 measurements were taken for each magnification and the average result was selected. The deviations between the measurements were less than 10% for each sample. It should be emphasized that a polished cross-section with a relatively constant value of SDAS was chosen, but due to the randomness of dendrite growth, some minor variations in S values are possible. Additionally, for the purpose of training, polished cross sections were chosen that contained defects. The defects could be roughly classified into five groups as shown in Figure 4: (a), (b)—air and shrinkage porosity defects; (c), (d)—scratches; (e)—blurred image; (c), (f), (g), (i)—different brightness and contrast of the image; (g), (h)—externally consolidated crystals (ECS). The presence of defects is beneficial to build the model's resistance to industry working conditions, as such defects are often seen in samples. About 10% of the dataset were images showing materials with some kind of defects. We should reiterate that images with this type of defect should not be used for SDAS predictions and are for training purposes only, as explained earlier. Finally, note that the Figures 3 and 4 contain images as supplied to the neural network. This is the reason for a lower standard of quality than usual in material science. However, the original higher quality images are available as supplementary materials to this paper (Figures S1 and S2).

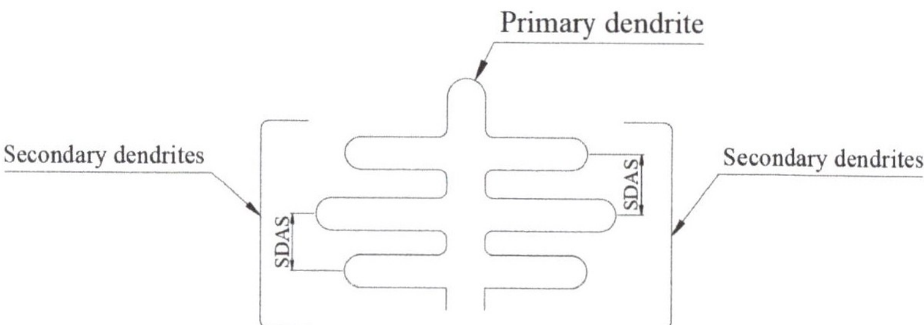

Figure 1. SDAS definition: the distance between two secondary dendrites.

The image dataset for training consists of 200×200 pixel images created from the original 2080×1544 pixel images, as shown in Figure 5. The original image was first split into 70 images with a pixel resolution of 208×211, which were then resized to 200×200 pixel. The S values and image count used for training, validation and testing are listed in

Table 2. The training set, validation set and test set were split in the ratio 60:20:20. The data instances were evenly distributed in terms of S values.

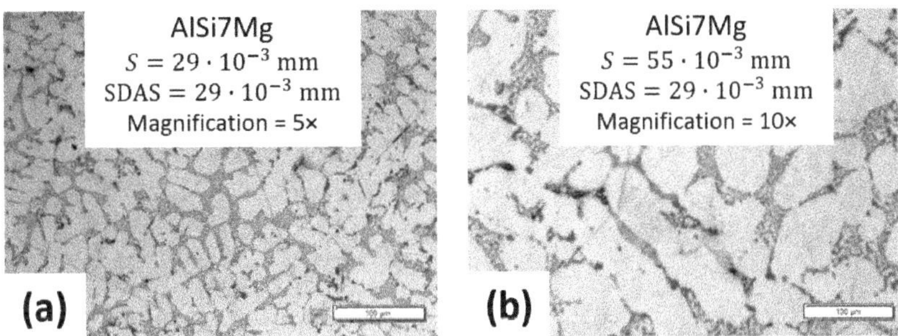

Figure 2. The procedure of deriving different S values using different magnifications on the microscope: (**a**) 5× magnification image; (**b**) 10× magnification image. The scale bar corresponds to S value.

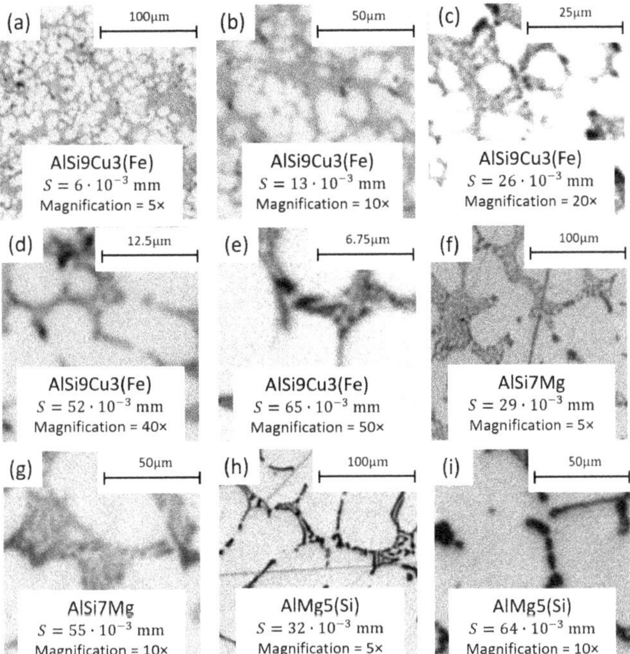

Figure 3. 200 × 200 pixel images used for training, their derived S values and alloy.

Table 1. Magnification factor F and appropriate pixel per micrometer value.

F	Magnification	Pixel/10^{-3} mm
1	5×	1.36
0.5	10×	2.72
0.25	20×	5.44
0.125	40×	10.88
0.1	50×	13.60

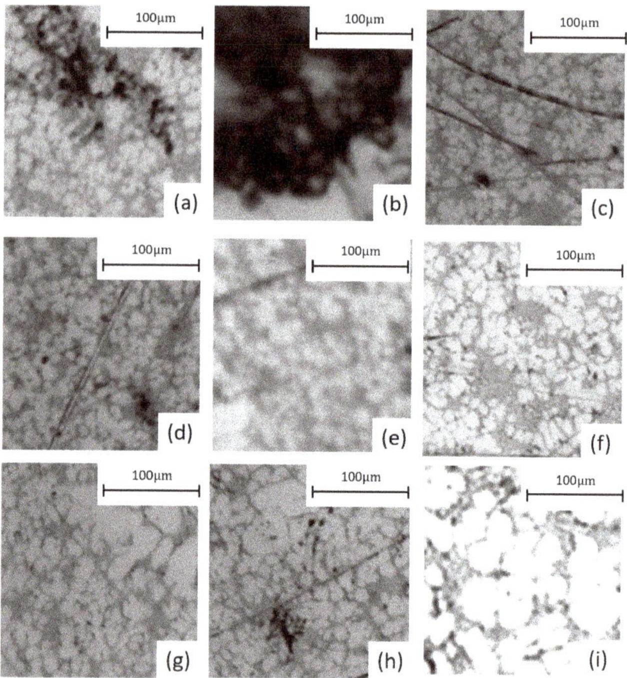

Figure 4. Images of different types of defects used for the training: (**a,b**)—air and shrinkage porosity defects; (**c,d**)—scratches; (**e**)—blurred image; (**c,f,g,i**)—different brightness and contrast of the image; (**g,h**)—ECS (some images are purportedly of lower quality).

Table 2. Images count per different material and different magnification.

Alloy	Magnification	S	F	S-SDAS/F	Images Count
AlSi9Cu3(Fe)	5×	6×10^{-3} mm	1	0	1076
AlSi9Cu3(Fe)	10×	13×10^{-3} mm	0.5	1	6820
AlSi9Cu3(Fe)	20×	26×10^{-3} mm	0.25	2	815
AlSi9Cu3(Fe)	40×	52×10^{-3} mm	0.125	4	1236
AlSi9Cu3(Fe)	50×	65×10^{-3} mm	0.1	5	1128
AlSi7Mg	5×	29×10^{-3} mm	1	0	1216
AlSi7Mg	10×	55×10^{-3} mm	0.5	3	1176
AlMg5(Si)	5×	32×10^{-3} mm	1	0	1032
AlMg5(Si)	10×	64×10^{-3} mm	0.5	0	1174

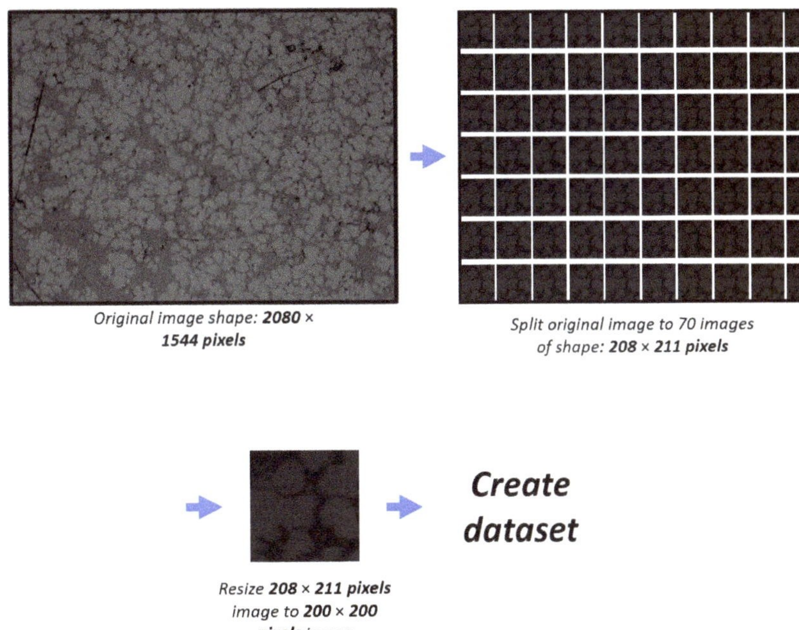

*Original image shape: **2080** × **1544** pixels*

*Split original image to 70 images of shape: **208** × **211** pixels*

*Resize **208** × **211** pixels image to **200** × **200** pixels image*

Create dataset

Figure 5. Schematic representation of dataset generation: the original image was first split into 70 images with a pixel resolution of 208 × 211, which were then resized to 200 × 200 pixel.

In an earlier approach [34], the images were preprocessed by the morphological transformation, Gaussian blur and Otsu's threshold [35] to reduce the number of training parameters. The color scale of the image was reduced from an eight-bit three-channel depth (i.e., full RGB) to a one-bit color depth (e.g., black or white). Unfortunately, the approach requires manual tuning of the hyperparameters for each SDAS value, making it impractical. For the present study, grayscale images were used to achieve fully automatic detection of SDAS using DL to avoid any manual operation. Nevertheless, the training parameters were reduced by a factor of three by using only one color channel instead of three.

3.3. Overview of the CNN Model

Predictive model at hand is based on the CNN architecture. It is assumed that other ML approaches would exhibit inferior behavior and therefore they were not considered. In the present case, the CNN model is based on a basic feedforward (sequential) network. The model consists of three 2D convolutional layer blocks. Each convolutional filter is followed by a Rectified Linear Unit (ReLU) activation, batch normalization and a max pooling layer. Filter sizes used were $(5,5) \times 32$, $(3,3) \times 32$ and $(2,2) \times 32$ for the first, second and third convolutional layers, respectively. Zero-padding was applied evenly across both dimensions to compensate for edges. The convolutions were performed with stride 1. Each max-pooling filter was of size $(2,2)$, and pooling was performed with stride 2. This resulted in the following intermediate activation maps: $200 \times 200 \times 1$ (input), $100 \times 100 \times 32$ (following the first convolutional block), $50 \times 50 \times 32$ (following the second convolutional block) and $25 \times 25 \times 32$ (following the third convolutional block). Activation of the final convolutional block was then flattened to create a meta-layer of 20,000 neurons, which was then followed by a fully-connected layer consisting of 64 neurons, followed by the ReLU activation function and batch normalization. Finally, this layer was then fully connected to the output layer, which contained a single linear neuron—because of the assumed regression operation. Thus, for a 2D input image of size $200 \times 200 \times 1$, a single real value—S value is obtained. The Adam optimizer was used, using a learning rate of 0.0005. The schematic CNN architecture is shown in Figure 6. The total number of parameters was 1,294,977. It was trained with the Keras library on an Intel® Core™ I7-4970 CPU running at 3.60 GHz using 8 parallel processing units. Training was completed in 5 h—for 30 epochs total. A batch size of 32 was used for the training.

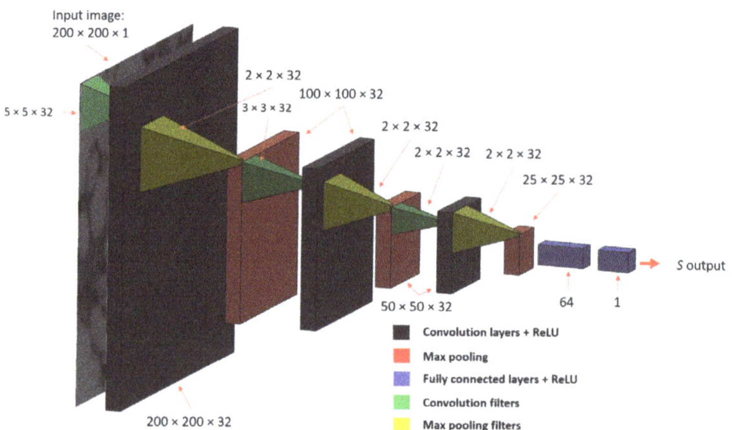

Figure 6. The CNN architecture used for estimating SDAS directly from microstructure image patches.

4. Results and Discussion

The CNN model described in the previous section showed very good performance, having the R^2 score of 91.5% on the test set. We performed an additional experiment to test the accuracy and practical usability of the model for possible industrial applications. The input image of 2080 × 1544 pixel is taken as a reference and split into 70 images of 200 × 200 pixel using the same approach as shown in Figure 5. The average result of 70 predictions is then captured while the prediction deviation and prediction error were also captured. We performed two mutually independent evaluation tests: one using materials that were used during the training, and another using materials that were not used during the training. It should be noted that individual images of both group of materials that were used for this test were not involved in the training, i.e., disjoint test subsets were used for both. The results are shown in Tables 3 and 4, while the images used for the prediction task are shown in Figures 7 and 8. S values were obtained using the same procedure as explained in the previous section.

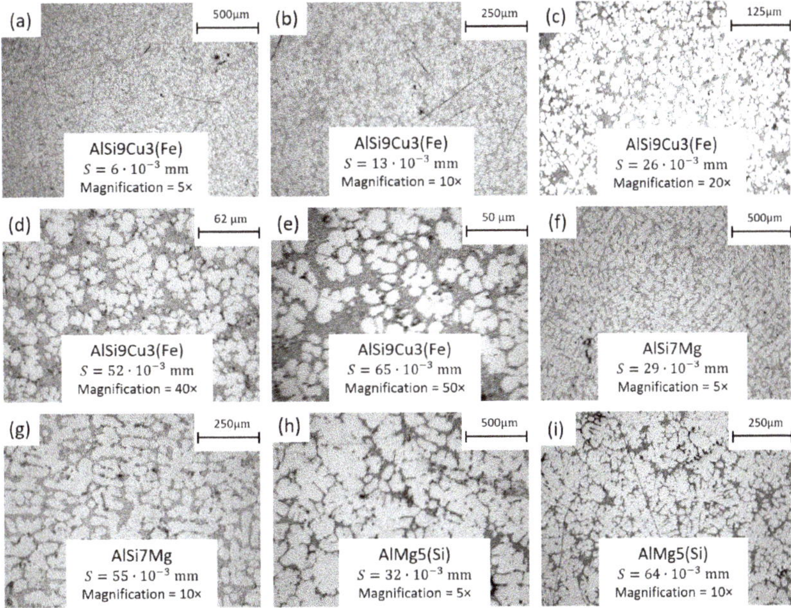

Figure 7. Images of materials used during the training: The images were split into 200 × 200 pixel images following the procedure in Figure 5 and used for the prediction task.

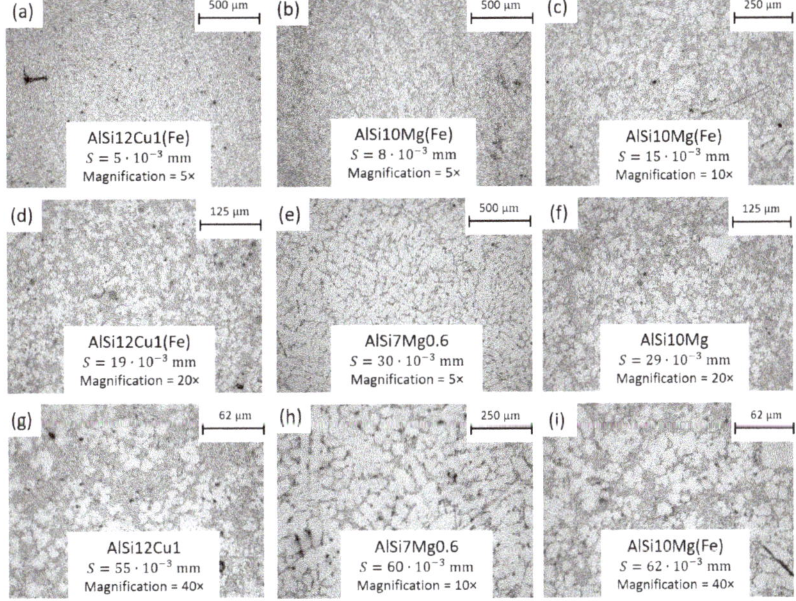

Figure 8. Images of materials that were not used during the training: The images were split into 200 × 200 pixel images following the procedure in Figure 5 and used for the prediction task.

Table 3. Results of prediction for materials used during training.

Material	S μm	Pred. S μm	Pred. Dev. μm	Pred. Err. %	Magnification
AlSi9Cu3(Fe)	6	4.8	1.2	20	5×
AlSi9Cu3(Fe)	13	13.6	0.6	4.6	10×
AlSi9Cu3(Fe)	27	27.7	0.7	2.6	20×
AlSi9Cu3(Fe)	52	52.9	0.9	1.7	40×
AlSi9Cu3(Fe)	65	64.9	0.1	0.15	50×
AlMg5(Si)	32	29.8	2.2	6.9	5×
AlMg5(Si)	64	62.6	1.4	2.2	10×
AlSi7Mg	29	23.0	6.0	20.7	5×
AlSi7Mg	55	55.9	0.9	1.6	10×

Average prediction deviation on 10× magnification: 0.76 μm
Average prediction error on 10× magnification: 2.15%

Table 4. Results of prediction for materials not used during training.

Material	S μm	Pred. S μm	Pred. Dev. μm	Pred. Err. %	Magnification
AlSi7Mg0.6	30	28.1	1.9	6.3	5×
AlSi7Mg0.6	60	58.1	1.9	3.1	10×
AlSi10Mg(Fe)	8	11.1	3.1	38.8	5×
AlSi10Mg(Fe)	15	23.9	8.9	59.3	10×
AlSi10Mg(Fe)	31	35.8	4.8	15.5	20×
AlSi10Mg(Fe)	62	52.7	9.3	15.0	40×
AlSi12Cu1(Fe)	5	13.9	8.9	178	5×
AlSi12Cu1(Fe)	19	26.1	7.1	37.4	20×
AlSi12Cu1(Fe)	40	41.4	1.4	3.5	40×

Average prediction deviation on max. mag.: 4.60 μm
Average prediction error on max. mag.: 7.2%

The predicted results are very accurate for the trained materials. Average prediction deviation was evaluated for 10× magnification, giving only $0.97 \cdot 10^{-3}$ mm. Moreover, the maximum prediction deviation is $6 \cdot 10^{-3}$ mm, while every other prediction deviation is less than $2.2 \cdot 10^{-3}$ mm. In Table 3, it can be observed that a higher prediction deviation as well as a higher prediction error correlates with a smaller magnification. This could be directly attributed to the resolution of the images. At higher magnification, a dendrite is described with more pixels, so it is expected that the predicted results are more accurate. For example, the highest magnification of 50× results in an error of only 0.1 μm. Hopefully, SDAS smaller than $5 \cdot 10^{-3}$ mm is rarely seen in industrial applications. However, even assuming the highest deviation, such SDAS prediction model performances are quite acceptable for industry. A model with such performance could even be used to determine the SDAS distribution on a single microstructure image and/or polished cross-section sample.

The prediction error for the predicted results was also recorded. The average prediction error for the trained materials at 10× magnification is very low −2.15% while the highest is 20.7%. Again, the highest prediction errors were associated with the lowest resolution for all alloys considered. Increasing the number of pixels that depicts SDAS through higher magnification effectively eliminates the accuracy problem. At the highest magnifications, the errors were in the range of 0.15–2.2%, as can be seen in Table 3. It is therefore recommended to use a magnification of at least 10×.

In addition, prediction accuracy was also tested with materials that were not used during training (Table 4). The physical SDAS values of AlSi10Mg(Fe), AlSi12Cu1(Fe) and AlSi7Mg0.6 were $8 \cdot 10^{-3}$ mm, $5 \cdot 10^{-3}$ mm and $30 \cdot 10^{-3}$ mm, respectively. The results shown in Table 4 indicate that the predictions are less accurate than for the trained materials. The prediction error is also slightly higher, while the average prediction deviation is $4.60 \cdot 10^{-3}$ mm for highest magnifications. For the AlSi10Mg(Fe) alloy, the highest error of 15% was obtained for $40\times$ magnification, while errors close to 3% were obtained for the other alloys. Note that a very high error (178%) was obtained for AlSi12Cu1(Fe) at low magnification, indicating that higher magnifications should again be preferred for the application of the CNN trained on other materials. This allows the error to be reduced to 3.5% for the $40\times$ magnification, which is in the same range as when the CNN is applied to trained materials, cf. Table 3. For general analysis (e.g., to determine whether the component was cast in sand or in permanent mold, or for general determination of solidification rate), the model could be used successfully in the foundry industry, ensuring that magnification greater than $10\times$ is used. As an example, even the deviation of the different measurement methods reported in [33] falls within the performance range of the model. The highest prediction deviation of $9.3 \cdot 10^{-3}$ mm was obtained for the alloy AlSi10Mg(Fe).

5. Conclusions

The present study confirms the hypothesis that SDAS can be determined directly from microstructure image data by the DL methodology. The model shows excellent performance on the materials used in the training, provided that the highest possible magnification is used. On the other hand, the model performs slightly worse for the materials not used in training. However, both variants could be used in any type of industry depending on the model accuracy. A model for materials used during training could even be used to determine the distribution of SDAS on a single microstructure image and/or polished cross-section sample. Furthermore, the model could still be applied for materials not used during training, although some caution should be exercised. Thus, from the discussion shown, it appears that the technique based on DL can predict SDAS with a high degree of certainty.

Since manual tuning of hyperparameter values is avoided in the present study, such an approach could be fully automated. Furthermore, this approach now has a potential application as part of Industry 4.0 for the automotive or other sector. Additionally, the methods of DL could also have a potential application for automated prediction of mechanical properties from an image of the microstructure. However, for this type of task, the link between SDAS and material properties should be part of the analysis.

Supplementary Materials: The following are available online at https://www.mdpi.com/article/10.3390/met11050756/s1, Figure S1: Images used for training (Figure 3, full resolution). Figure S2: Images of different types of defects used for the training (Figure 4, full resolution).

Author Contributions: Conceptualization, methodology, software, validation, investigation, writing—review and editing: F.N., I.Š. and M.Č. All the authors contributed equally to this work. All authors have read and agreed to the published version of the manuscript.

Funding: This research received no external funding.

Data Availability Statement: Data required to reproduce these findings are available for download from: http://doi.org/10.17632/jkp57kbzxn.1, accessed on 31 March 2021. The following dataset consists of figures which are used during the training of the convolutional neural network as well as saved weights of the convolutional neural network after the training.

Acknowledgments: This work has been partially supported by the University of Rijeka under the project number uniri-tehnic-18-37. Polished samples were purchased from the University of Ljubljana: Faculty of Natural Sciences and Engineering. Additional samples were donated by TC Livarstvo d.o.o, and CIMOS d.d. Automotive Industry. This support is gratefully acknowledged.

Conflicts of Interest: The authors declare no conflict of interest.

References

1. Vandersluis, E.; Ravindran, C. Relationships between solidification parameters in A319 aluminum alloy. *J. Mater. Eng. Perform.* **2018**, *27*, 1109–1121. [CrossRef]
2. Vandersluis, E.; Ravindran, C. Influence of solidification rate on the microstructure, mechanical properties, and thermal conductivity of cast A319 Al alloy. *J. Mater. Sci.* **2019**, *54*, 4325–4339. [CrossRef]
3. Djurdjevič, M.; Grzinčič, M. The effect of major alloying elements on the size of the secondary dendrite arm spacing in the as-cast Al-Si-Cu alloys. *Arch. Foundry Eng.* **2012**, *12*, 19–24. [CrossRef]
4. Brusethaug, S.; Langsrud, Y. Aluminum properties, a model for calculating mechanical properties in AlSiMgFe-foundry alloys. *Metall. Sci. Tecnol.* **2000**, *18*, 3–7.
5. Shabestari, S.; Shahri, F. Influence of modification, solidification conditions and heat treatment on the microstructure and mechanical properties of A356 aluminum alloy. *J. Mater. Sci.* **2004**, *39*, 2023–2032. [CrossRef]
6. Kabir, M.S.; Ashrafi, A.; Minhaj, T.I.; Islam, M.M. Effect of foundry variables on the casting quality of as-cast LM25 aluminium alloy. *Int. J. Eng. Adv. Technol.* **2014**, *3*, 115–120.
7. Seifeddine, S.; Wessen, M.; Svensson, I. Use of simulation to predict microstructure and mechanical properties in an as-cast aluminium cylinder head comparison-with experiments. *Metall. Sci. Tecnol.* **2006**, *24*, 7.
8. Morri, A. Empirical models of mechanical behaviour of Al-Si-Mg cast alloys for high performance engine applications. *Metall. Sci. Technol.* **2010**, *28*, 25–32.
9. Grosselle, F.; Timelli, G.; Bonollo, F.; Della Corte, E. Correlation between microstructure and mechanical properties of Al-Si cast alloys. *La Metallurgia Italiana* **2009**, *27*, 25–32.
10. Qi, M.; Li, J.; Kang, Y. Correlation between segregation behavior and wall thickness in a rheological high pressure die-casting AC46000 aluminum alloy. *J. Mater. Res. Technol.* **2019**, *8*, 3565–3579. [CrossRef]
11. Singh, S.; Bhadeshia, H.; MacKay, D.; Carey, H.; Martin, I. Neural network analysis of steel plate processing. *Ironmak. Steelmak.* **1998**, *25*, 355–365.
12. Hancheng, Q.; Bocai, X.; Shangzheng, L.; Fagen, W. Fuzzy neural network modeling of material properties. *J. Mater. Process. Technol.* **2002**, *122*, 196–200. [CrossRef]
13. Agrawal, A.; Deshpande, P.D.; Cecen, A.; Basavarsu, G.P.; Choudhary, A.N.; Kalidindi, S.R. Exploration of data science techniques to predict fatigue strength of steel from composition and processing parameters. *Integr. Mater. Manuf. Innov.* **2014**, *3*, 8. [CrossRef]
14. Liu, R.; Kumar, A.; Chen, Z.; Agrawal, A.; Sundararaghavan, V.; Choudhary, A. A predictive machine learning approach for microstructure optimization and materials design. *Sci. Rep.* **2015**, *5*, 11551. [CrossRef]
15. Yang, X.W.; Zhu, J.C.; Nong, Z.S.; Dong, H.; Lai, Z.H.; Ying, L.; Liu, F.W. Prediction of mechanical properties of A357 alloy using artificial neural network. *Trans. Nonferrous Met. Soc. China* **2013**, *23*, 788–795. [CrossRef]
16. Santos, I.; Nieves, J.; Penya, Y.K.; Bringas, P.G. Machine-learning-based mechanical properties prediction in foundry production. In Proceedings of the 2009 ICCAS-SICE, Fukuoka City, Japan, 18–21 August 2009; pp. 4536–4541.
17. Liao, H.; Zhao, B.; Suo, X.; Wang, Q. Prediction models for macro shrinkage of aluminum alloys based on machine learning algorithms. *Mater. Today Commun.* **2019**, *21*, 100715. [CrossRef]
18. Agrawal, A.; Choudhary, A. Deep materials informatics: Applications of deep learning in materials science. *MRS Commun.* **2019**, *9*, 779–792. [CrossRef]
19. Herriott, C.; Spear, A.D. Predicting microstructure-dependent mechanical properties in additively manufactured metals with machine-and deep-learning methods. *Comput. Mater. Sci.* **2020**, *175*, 109599. [CrossRef]
20. Ferguson, M.K.; Ronay, A.; Lee, Y.T.T.; Law, K.H. Detection and segmentation of manufacturing defects with convolutional neural networks and transfer learning. *Smart Sustain. Manuf. Syst.* **2018**, *2*, 1–42. [CrossRef]
21. Hanumantha Rao, D.; Tagore, G.; Ranga Janardhana, G. Evolution of Artificial Neural Network (ANN) model for predicting secondary dendrite arm spacing in aluminium alloy casting. *J. Braz. Soc. Mech. Sci. Eng.* **2010**, *32*, 276–281. [CrossRef]
22. Wei, J.; Chu, X.; Sun, X.Y.; Xu, K.; Deng, H.X.; Chen, J.; Wei, Z.; Lei, M. Machine learning in materials science. *InfoMat* **2019**, *1*, 338–358. [CrossRef]
23. Azimi, S.M.; Britz, D.; Engstler, M.; Fritz, M.; Mücklich, F. Advanced steel microstructural classification by deep learning methods. *Sci. Rep.* **2018**, *8*, 2128. [CrossRef]
24. Li, X.; Zhang, Y.; Zhao, H.; Burkhart, C.; Brinson, L.C.; Chen, W. A transfer learning approach for microstructure reconstruction and structure-property predictions. *Sci. Rep.* **2018**, *8*, 13461. [CrossRef]
25. Mery, D. Aluminum Casting Inspection Using Deep Learning: A Method Based on Convolutional Neural Networks. *J. Nondestruct. Eval.* **2020**, *39*, 12. [CrossRef]
26. DeCost, B.L.; Lei, B.; Francis, T.; Holm, E.A. High throughput quantitative metallography for complex microstructures using deep learning: A case study in ultrahigh carbon steel. *arXiv* **2018**, arXiv:1805.08693.
27. Chowdhury, A.; Kautz, E.; Yener, B.; Lewis, D. Image driven machine learning methods for microstructure recognition. *Comput. Mater. Sci.* **2016**, *123*, 176–187. [CrossRef]
28. Exl, L.; Fischbacher, J.; Kovacs, A.; Oezelt, H.; Gusenbauer, M.; Yokota, K.; Shoji, T.; Hrkac, G.; Schrefl, T. Magnetic microstructure machine learning analysis. *J. Phys. Mater.* **2018**, *2*, 014001. [CrossRef]

29. Yucel, B.; Yucel, S.; Ray, A.; Duprez, L.; Kalidindi, S.R. Mining the Correlations Between Optical Micrographs and Mechanical Properties of Cold-Rolled HSLA Steels Using Machine Learning Approaches. *Integr. Mater. Manuf. Innov.* **2020**, *9*, 240–256. [CrossRef]
30. Pokuri, B.S.S.; Ghosal, S.; Kokate, A.; Sarkar, S.; Ganapathysubramanian, B. Interpretable deep learning for guided microstructure-property explorations in photovoltaics. *NPJ Comput. Mater.* **2019**, *5*, 95. [CrossRef]
31. DeCost, B.L.; Holm, E.A. A computer vision approach for automated analysis and classification of microstructural image data. *Comput. Mater. Sci.* **2015**, *110*, 126–133. [CrossRef]
32. *DIN EN 1706: Aluminium und Aluminiumlegierungen-Gussstücke-Chemische Zusammensetzung und mechanische Eigenschaften*; Beuth: Berlin, Germany, 1998.
33. Vandersluis, E.; Ravindran, C. Comparison of measurement methods for secondary dendrite arm spacing. *Metallogr. Microstruct. Anal.* **2017**, *6*, 89–94. [CrossRef]
34. Nikolic, F.; Štajduhar, I.; Čanađija, M. Aluminum microstructure inspection using deep learning: A convolutional neural network approach toward secondary dendrite arm spacing determination. In *4th Edition of My First Conference*; University of Rijeka, Faculty of Engineering: Rijeka, Croatia, 2020.
35. Alexander Mordvintsev, A.K.R. Image Processing in OpenCV. 2013. Available online: https://opencv-python-tutroals.readthedocs.io/en/latest/py_tutorials/py_imgproc/py_table_of_contents_imgproc/py_table_of_contents_imgproc.html (accessed on 5 September 2020).

Article

Influence of Additive Manufactured Stainless Steel Tool Electrode on Machinability of Beta Titanium Alloy

Ragavanantham Shanmugam [1,*], Monsuru Ramoni [2,*], Geethapriyan Thangamani [3,*] and Muthuramalingam Thangaraj [4,*]

1   Advanced Manufacturing Technology, School of Engineering, Math and Technology, Navajo Technical University, Crownpoint, NM 87313, USA
2   Industrial Engineering, School of Engineering, Math and Technology, Navajo Technical University, Crownpoint, NM 87313, USA
3   Department of Mechanical Engineering, SRM Institute of Science and Technology, SRM Nagar, Kattankulathur 603203, India
4   Department of Mechatronics Engineering, SRM Institute of Science and Technology, SRM Nagar, Kattankulathur 603203, India
*   Correspondence: rags@navajotech.edu (R.S.); mramoni@navajotech.edu (M.R.); geethapt@srmist.edu.in (G.T.); muthurat@srmist.edu.in (M.T.)

Abstract: Additive manufacturing technology provides a gateway to completely new horizons for producing a wide range of components, such as manufacturing, medicine, aerospace, automotive, and space explorations, especially in non-conventional manufacturing processes. The present study analyzes the influence of the additive manufactured tool in electrochemical micromachining (ECMM) on machining beta titanium alloy. The influence of different machining parameters, such as applied voltage, electrolytic concentration, and duty ratio on material removal rate (MRR), overcut, and circularity was also analyzed. It was inferred that the additive manufactured tool can produce better circularity and overcut than a bare tool due to its higher corrosion resistance and localization effect. The additive manufactured tool can remove more material owing to its strong atomic bond of metals and higher electrical conductivity.

Keywords: ECMM; beta titanium alloy; electrolyte; stainless steel; additive tool

Citation: Shanmugam, R.; Ramoni, M.; Thangamani, G.; Thangaraj, M. Influence of Additive Manufactured Stainless Steel Tool Electrode on Machinability of Beta Titanium Alloy. *Metals* **2021**, *11*, 778. https://doi.org/10.3390/met11050778

Academic Editor: Claudio Testani

Received: 6 April 2021
Accepted: 7 May 2021
Published: 10 May 2021

Publisher's Note: MDPI stays neutral with regard to jurisdictional claims in published maps and institutional affiliations.

1. Introduction

The conventional manufacturing technique has limitations for producing complex shapes, micro-holes, micro slots, as well as more tool wear and limited dimensional accuracy and precision [1,2]. Nontraditional manufacturing processes are used to overcome those problems [3,4]. The complex geometry ability of additive manufacturing offers an opportunity to produce complex geometrical tools for non-conventional manufacturing processes, especially electrochemical machining (ECM) processes. It is important to study the influence of the additive manufacturing (AM) tool in electrochemical micro-machining (ECMM) with related to titanium alloy. Such analysis has the potential to revolutionize the manufacturing processes of tools for non-conventional manufacturing applications. The various other non-conventional manufacturing techniques have a heat affected zone on the work piece. However ECM process does not produce any such heat affected zone (HAZ). It makes ECMM a great manufacturing process compared to other unconventional techniques. In this process, the workpiece acts as an anode to achieve the controlled removal of metal by anodic dissolution. The tool electrode acts as cathode while electrolyte flows through the inter-electrode gap. The shape of the tool is generally a replica of the shape to be machined on the workpiece [5]. Electrochemical machining is based on Faraday's laws of electrolysis and Ohm's law [6]. Two electrodes are arranged in an electrolytic solution. When DC power is supplied across these electrodes, the metal is eroded from the anode and deposited on the cathode. ECMM process is the replica of electrochemical plating technique. In the

Metals **2021**, *11*, 778. https://doi.org/10.3390/met11050778

ECMM process, the material is removed at an atomic level through anodic dissolution, when high current is conducted at a relatively low potential difference through a specially shaped tool. As the ECMM process provides high surface finish, high accurate and stress-free products, it is widely used in manufacturing of turbine blades, high compression engines, artillery projectiles and parts for electronics and medical industries [7]. ECMM is performed for very small size products with a better controlled ECM process. It is used to produce micro-holes and micro-slots in the material with very high precision and quality [8]. Theoretically, in the micro machining process, the micro-holes generated within the dimension of 1 µm to 999 µm [9,10]. The ECMM has been referred for many of the micromachining because it removes the materials (conductive materials) regardless of their hardness and toughness [11]. Additive manufacturing (AM) is the process of generating a product layer by layer from a 3D model. It enables manufacturing of complex geometry with fewer processing step and minimum waste; it reduces time and cost for manufacturing products. It is now applied in the aerospace, medical implant, and electronics fields. Many research works have been done on the ECM process based on its process parameters, such as electrolytes, tool shape, materials, and workpiece. Some other studies on coated tools have also been performed with various tool electrodes [12]. This study is about machining of beta titanium alloy as a workpiece of ECMM process and the influence of additively manufactured tool over the process, which are yet to be studied. The machining performance on titanium alloy specimens using the ECM process was improved using an optimal combination of electrolytes and their concentration during the drilling process [13]. The nickel-based alloys can be precisely machined using ECM process by choosing optimal process parameters combination [14]. The influence of the tool coating was also analyzed on machining titanium alloy [15]. The surface morphology of the machined specimens should be improved as much as possible [16,17].

From the referenced literature, it has been understood that very little attention has been given to the analysis of the influence of the additively manufactured tool electrode on quality measures using the ECMM process. The objective of the study is to study the influence of process parameters and additively manufactured copper tool electrode on material removal rate (MRR), overcut (OC), and circularity (CY) while machining Titanium (Ti-6Al-4V) alloy in ECMM process. The micro structural analysis was also carried out to understand the variation of grains around the machined area of each specimen.

2. Materials and Methods

2.1. Electro. Chemical Micro Machining Arrangement

The ECMM setup has a machining unit, micro electrode feeding system (lead screw mechanism), electrolytic tank, and DC power supply system, as shown in Figure 1. The electrolyte bath is connected to a pump and a filter. The mechanical machining unit consists of a work holding device, micro-tool feeding system and machining chamber. The micro-tool feed movement has been achieved through the lead screw mechanism. The lead screw of the tool movement has been rotated with the help of a stepper motor, whereas the tool movement can be controlled manually. A DC power supply of 0 V–30 V and 0 A–2 A is inbuilt with the ability to control the voltage, current, and duty cycle. The experimental factors have been selected based on the influence of the process parameters, which affect the machining rate and shape accuracy of the material. The ECMM setup consists of a power input, control unit to set the process parameters, electrolyte supply system and a machining chamber which contains tool and the workpiece along with the electrolyte [18]. Titanium (Ti-6Al-4V) alloy with the size of (50 mm × 50 mm × 3 mm) was chosen as a workpiece, due to its importance in manufacturing industries.

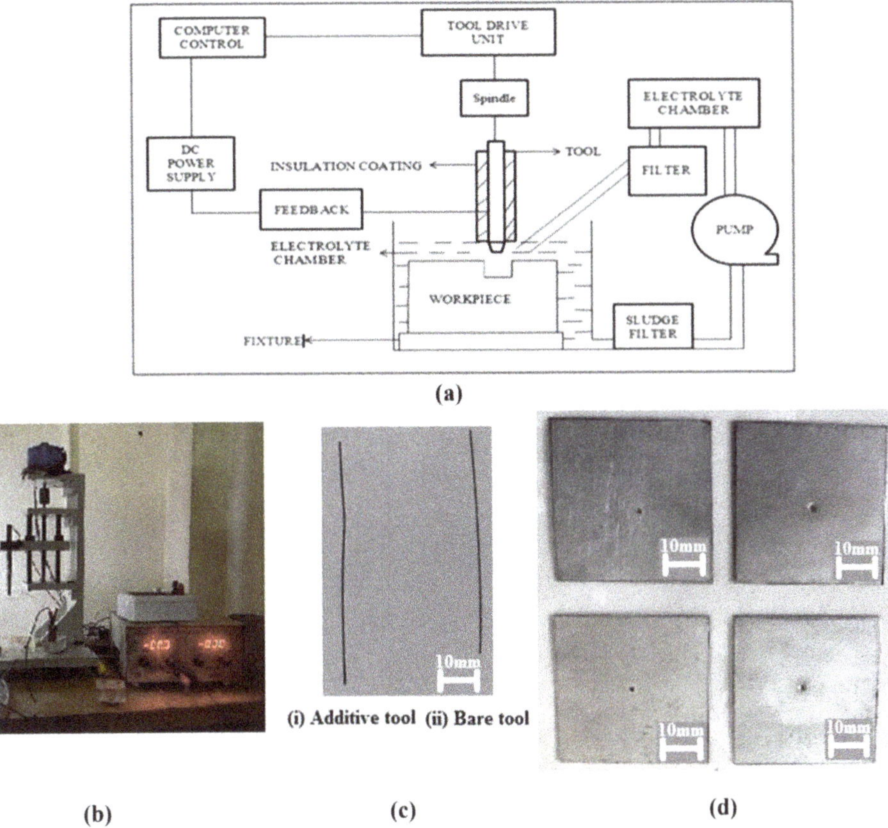

Figure 1. Experimental setup of ECMM with machined specimens and tool electrodes. (**a**) Schematic ECM, (**b**) ECM machine, (**c**) Tool electrodes, (**d**) Machined specimens.

2.2. Selection of Process Variables

This research work is focused on finding the machining performance of additively manufactured tool while machining beta titanium alloy by ECMM through calculating the selected output parameters, such as *MRR* and surface accuracy. The Stainless Steel 316L tool electrode in the shape of a sharp pencil hat is 30 mm long and 2 mm thick (0.4 mm thick from tip) has been used as a tool electrode in the ECMM process. The tool electrodes were manufactured using a conventional machining process (bare tool) and additive manufacturing (AM tool) process, as shown in Figure 1. The presence of a small volume of silicon and copper in the austenitic stainless steel containing molybdenum can improve the corrosion resistance by improving oxidation resistance. The additively manufactured stainless steel has almost the same properties compared to conventional available bare stainless steel. However, the AM tool electrode has better uniformity in composition and dimensional accuracy. The tool electrode was manufactured using Direct Metal Laser Sintering (DMLS) as the additive manufacturing technique. In this approach, the metal powders were utilized to build a platform continuously. The powder particles were sintered layer by layer using a laser beam. The shape of the job could be determined based on a CAM model. Then the specimens are formed into the required shape with the wire electrical discharge machining (WEDM) process [18]. Beta titanium alloy has been chosen as the work piece material of size of 40 mm × 20 mm × 1.7 mm. Table 1 shows the composition of the work piece and tool electrode. The importance of titanium material has been increasing in the field of biomedical because it fulfills the requirements over other materials with its properties, and also the titanium has wide applications in the area of aircraft, jet engines, racing cars, chemical, petrochemical, and marine components, and submarine hulls [19–22]. The 1 mol/L of Sodium Nitrate and 1 mol/L of Sodium Nitrate with 0.02 mol/L of Sodium Citrate were chosen as electrolyte, as shown in Table 2 [23–25]. In the present study, applied DC voltage, electrolytic concentration and duty cycle were selected as process parameters. The voltage was chosen as 15 V and 17 V with the Duty cycle of 50% and 66%. The manual Micro-tool feed rate of 1.6μm/min was chosen for this experiment [20,23]. During the machining process, the parameters were set to require based on a trial of an orthogonal array. The machining was done for making blind hole.

Table 1. Chemical Composition of workpiece and tool electrode.

Beta Titanium Alloy Workpiece	
Elements	**Composition (%)**
Iron (Fe)	0.23
Aluminum (Al)	3.01
Vanadium (V)	2.19
Titanium (Ti)	94.65
Stainless Steel 316l Tool electrode	
Carbon	0.03
Manganese	2.00
Phosphorus	0.045
Sulfur	0.03
Silicon	0.75
Chromium	18.00
Nickel	14.00
Molybdenum	3.00
Nitrogen	0.10
Iron	62.045

Table 2. Selection of Process parameters and its levels.

Input Parameters	Level 1	Level 2
Applied voltage (V)	15	17
Electrolytic concentration (mol/L)	1—NaNO$_3$	1—NaNO$_3$
	0—Sodium Citrate	0.02—Sodium Citrate
Duty cycle (%)	50	66

2.3. Selection of Performance Measures

Material removal rate (*MRR*) has been obtained by taking ratio between per machining time and weight difference of workpiece before and after machining. The values of the *MRR* were computed in terms of (g/h) using the following Equation (1).

$$MRR = \frac{W_b - W_a}{T} \tag{1}$$

where, W_b = Pre-Weight of the piece (g); W_a = After-Weight of the piece (g); T = Machining time (hours). The circularity indicates the dimensional accuracy which provides the deviation from the required shape. The value should be less as much as possible and denoted in µm. It was measured based on the following Equation (2) [17].

$$\text{Circularity} = D_{\text{MAJOR}} - D_{\text{MINOR}} \tag{2}$$

where, D_{MAJOR} = Major Diameter of Micro-hole (µm), D_{MINOR} = Minor Diameter of Micro-hole (µm). The overcut indicates the deviation from the required width of the cut. The value should be less as much as possible and denoted in µm. The difference between the cross section of electrode tip and micro-hole is known as overcut. It has been calculated by finding the difference between the radius of the tool and micro-hole as mentioned Equation (3). The tool giving lower value for overcut is superior.

$$\text{Overcut} = \frac{D_{\text{H}} - D_{\text{T}}}{2} \tag{3}$$

where, D_{H} = Mean Diameter of Micro-hole (µm); D_{T} = Diameter of Tool electrode Tip (µm). The profile projector was used to measure the dimension of tools. The shape accuracy and effect on the tool can be seen by optical microscope. The dimensions of the tool electrode were measured before and after machining using a profile projector. The Minitab package software was used for obtaining the optimal process parameters by analyzing the output characteristics data.

3. Results and Discussion

The blind hole machining operations were performed on titanium alloys specimens, as shown in Figure 1. All the experimental trials have been conducted with two times, and the average values were considered as final value to increase the measurement accuracy. The effects of AM tool and bare tool on performance measures while machining beta titanium alloy specimens under various process parameters like voltage, concentration and duty cycle using the ECMM process were compared and analyzed.

3.1. Influence of Additive Manufactured Tool Electrode on MRR

The significance of process parameters was investigated using Minitab 17 version software package. The most influencing ECMM variables can be indicated by the deviation from the mean line. It was inferred that the electrolyte concentration has the most dominating factor than others, such as applied voltage and duty cycle. The *MRR* was increased with a higher concentration of electrolytes due to this higher number of free ions and increasing current flow between the electrodes, as shown in Figure 2 [20,23]. *MRR* is directly proportional to voltage and duty cycle, but inversely proportional to electrolyte concentration. The additive tool has a more dominating effect on *MRR* because as by Ohm's law, voltage and current are directly related to *MRR*, as shown in Figure 3. The error bar also indicated the lower repeatability error and standard deviation. The voltage can increase the current in the electrolytic cell of the process, which result in higher *MRR* in the ECMM process [26,27]. Hence, the optimum voltage should be kept as high as possible. From the Table 3, it was inferred that additive manufactured tool could provide better *MRR* than bare tool owing to the high uniformity of composition in additively manufactured tool as compared to conventionally manufactured bare tool. The study was also performed to find the optimal parameters and investigate the effects on the performance measures. It was found that higher voltage and duty cycle could provide higher material removal rate with better dimensional accuracy, owing to the ability of producing better anodic dissolution. It forms strong atomic bonds of metals, which increases electrical conductivity of the material to increase *MRR* during the machining process in ECMM [28].

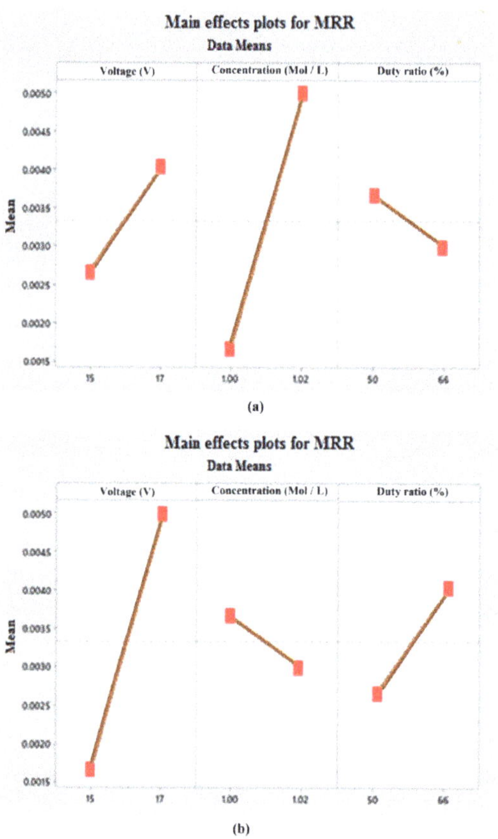

Figure 2. Contribution of process parameter on MRR (**a**) bare tool (**b**) additive tool.

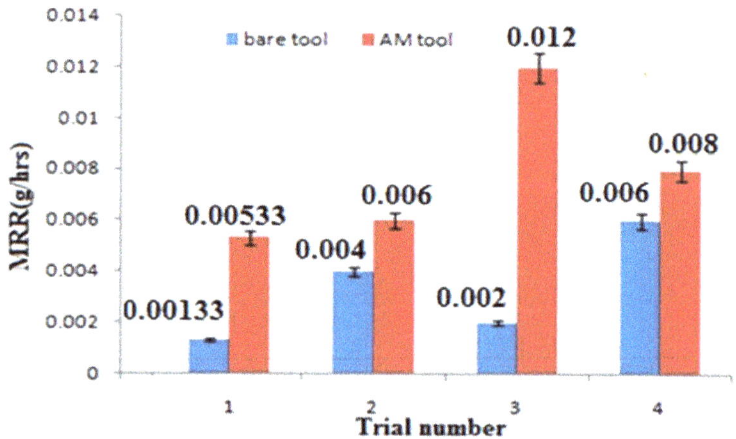

Figure 3. Comparison graph for bare tool and additive tool on *MRR*.

Table 3. Comparison between Bare Tool and Additive Tool for *MRR*.

S.No.	Voltage (V)	Concentration (Mol/L)		Duty Ratio (%)	MRR (g/hrs)	
		Sodium Nitrate	Sodium Citrate		Bare Tool	Additive Tool
1	15	1	0	50	0.00133	0.00533
2	15	1	0.02	66	0.004	0.006
3	17	1	0	66	0.002	0.012
4	17	1	0.02	50	0.006	0.008

3.2. Influence of Additive Manufactured Tool Electrode on Circularity and Overcut

The quality of the micro-hole produced by ECMM process on titanium specimens by AM tool and bare tool electrodes. It can be measured by capturing circularity and overcut of the workpiece under various process parameters combinations. The optical microscope (SVI107 manufactured by SIPCON, India) was used examine the quality of micro-hole by measure the value of circularity and overcut. The scanning electron microscope (SEM S-3400N manufactured by Hitachi, Japan) has been used to understand surface morphology in the present study.

Figure 4 shows the SEM image of a micro-hole on a beta titanium alloy workpiece machined using a SS316L bare tool in ECMM. The circle in the figure shows the micro-holes in the machined region of the workpiece. The workpiece was machined by bare tool under (15 V voltage, 1 mol/L Sodium Nitrate, and 50% duty cycle) and (15 V voltage, 1 mol/L Sodium Nitrate with 0.002 mol/L Sodium Citrate, and 66% duty cycle) for SEM analysis. It was observed that higher duty cycle increases the number of free ions for the electrolyte. This could increase *MRR* by increasing the current in the process owing to a higher duty cycle. It could increase the quality of micro-holes by producing larger and concentrated craters in the bare tool [29,30].

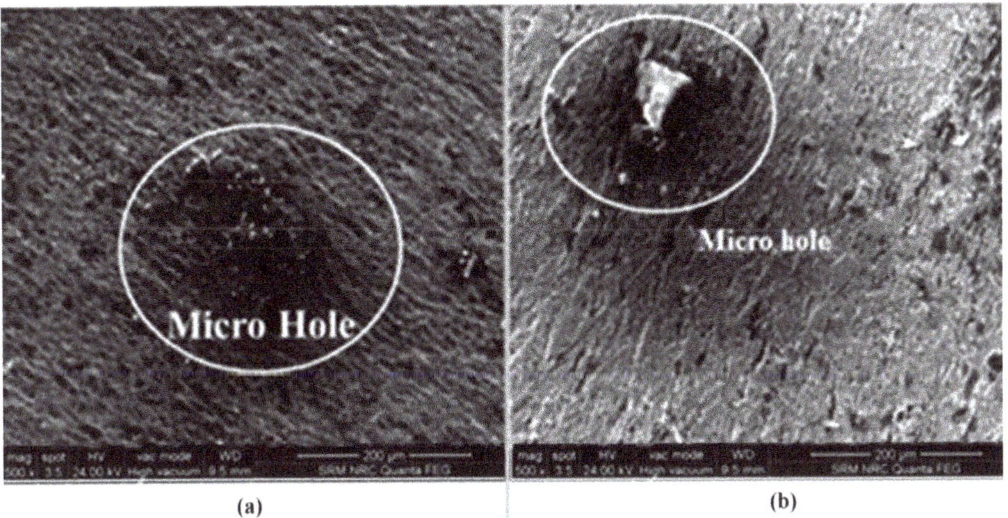

Figure 4. SEM image of machined workpiece surface by bare tool (**a**) 15 V, 1 mol/L Sodium Nitrate, 50% duty cycle (**b**) 15 V, 1 mol/L Sodium Nitrate, 66% duty cycle.

Figure 5 shows the SEM image of micro-hole on beta titanium alloy workpiece machined using the additive tool of SS316L in the ECMM process. The circle in the figure shows the micro-holes in the machined region of the workpiece. The workpiece machined by bare tool under (15 V voltage, 1 mol/L Sodium Nitrate and 50% duty cycle) and (15 V voltage, 1 mol/L Sodium Nitrate with 0.002 mol/L Sodium Citrate and 66% duty cycle) for SEM analysis. The higher duty cycle could produce larger and more concentrated craters in the ECMM process. The surface accuracy was also observed in better form due to the higher particle uniformity and corrosion resistance, which could reduce the unwanted material removal on the machined surface in the ECMM process. It could result in compromising the quality of micro-holes on the machined specimens.

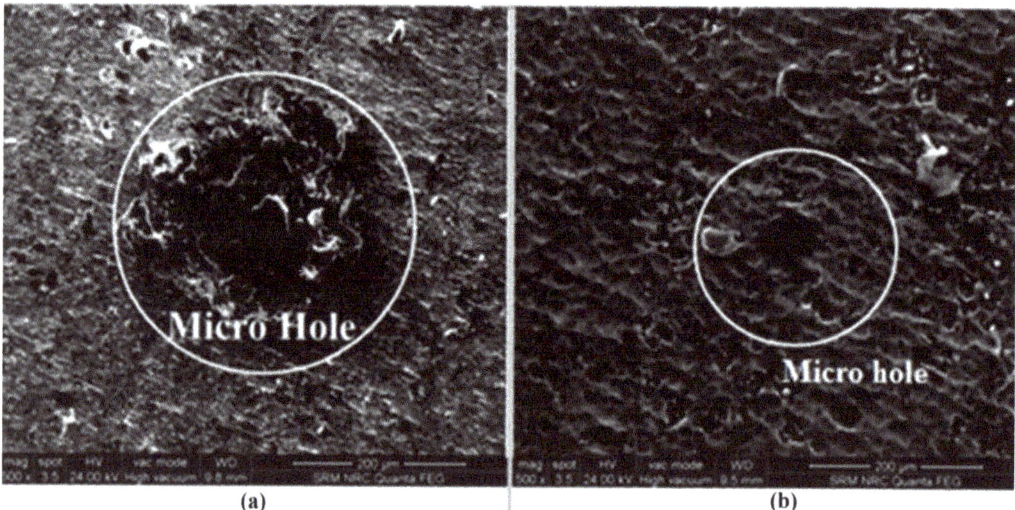

Figure 5. SEM image of machined workpiece surface by additive tool (a) 15 V, 1 mol/L Sodium Nitrate, 50% duty cycle (b) 15 V, 1 mol/L Sodium Nitrate, 66% duty cycle.

Table 4 shows that better circularity was observed with the additive tool than the bare tool. The dimensional exactness of the tool electrode demonstrates the dimensional accuracy over the machined specimens. Hence, the circularity of specimens was observed as better with the additive tool owing to the elemental exactness of the tool electrode, as shown in Figure 6. The reason behind the better dimensional accuracy of additive manufacturing could be due to the layer by layer manufacturing technique. It could avoid any shrinkage of metal during manufacturing.

Table 4. Comparison table between Bare Tool and Additive Tool for Circularity.

Trials	Circularity by Bare Tool (μm)	Circularity by Additive Tool (μm)
1	67.19	17.53
2	40.21	0.59
3	2.39	0.87
4	19.43	0.58

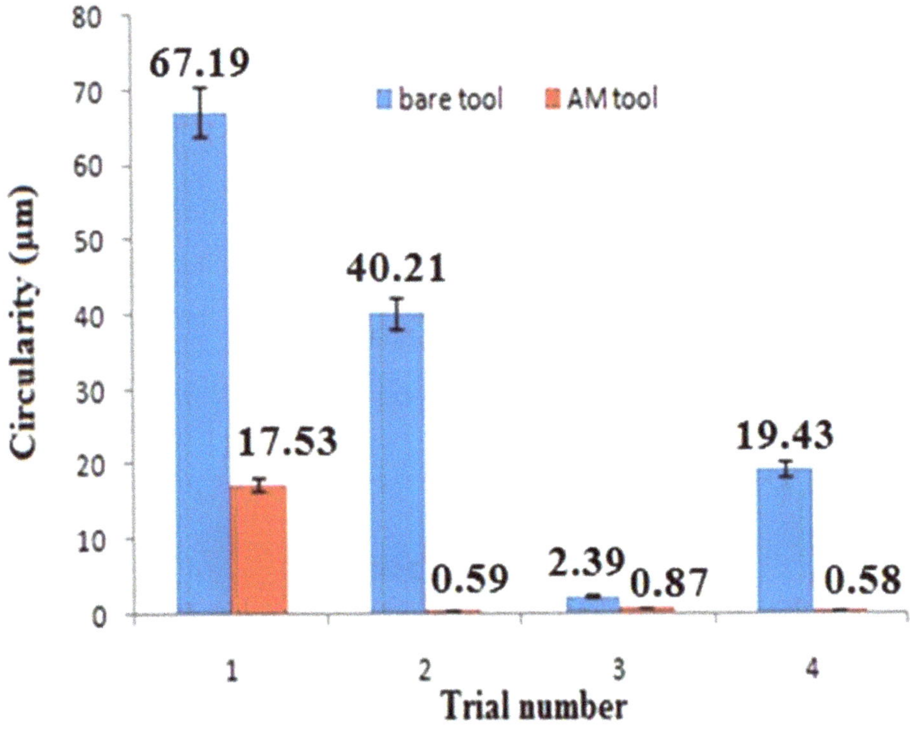

Figure 6. Comparison graph for bare tool and additive tool on circularity.

Table 5 shows that a lower overcut was observed with the additive tool than with the bare tool. The lower overcut of specimens was observed with additive tool due to the layer by layer additive manufacturing of the tool electrode, as shown in Figure 7. It could be due to the high localization effect and minimization of stray current provided by additive manufacturing tool in the ECMM process.

Table 5. Comparison table between Bare Tool and Additive Tool for Overcut.

Trials	Overcut by Bare Tool (µm)	Overcut by Additive Tool (µm)
1	25.5525	17.6575
2	21.29	16.4875
3	5.4175	4.7375
4	18.2725	9.99

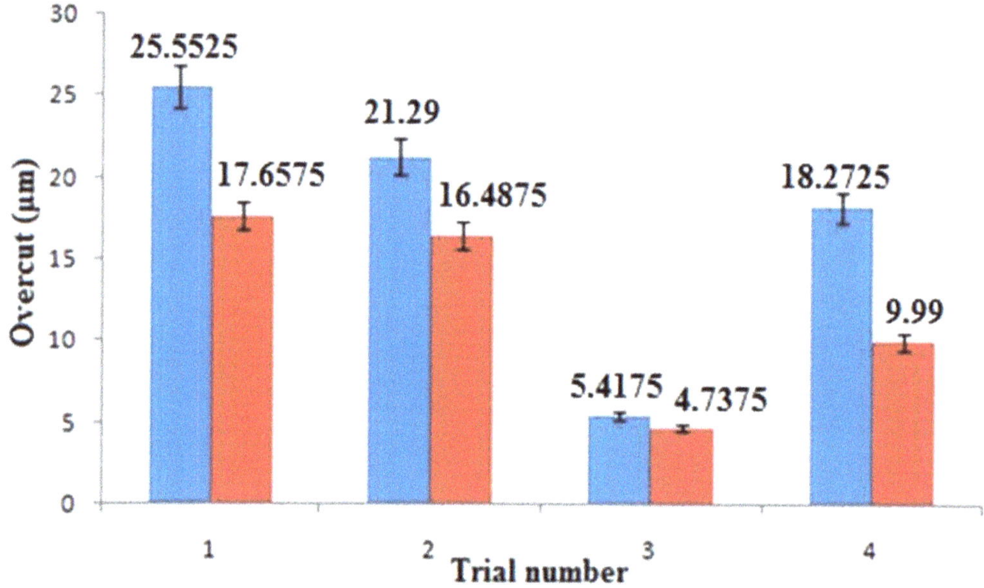

Figure 7. Comparison graph for bare tool and additive tool on overcut.

3.3. Surface Morphology Analysis with Additive Manufactured Tool Electrode

The corrosion can be seen in the tool at the tool tip and tool surface with a bare tool electrode as shown in Figure 8. The circle indicated the corrosion on the tool after the machining process. The corrosion in the tool is due to pitting corrosion, as the corrosion is not uniform throughout the tool and forming pits on the tool surface. Tool surface topology shows that the influence of the machining process over the tool electrode surface. After machining process, more craters were observed over the machined surface due to material removal in the ECMM process. It is due to defects that occurred during conventional manufacturing of the tool [31]. Figure 8 shows the corrosion at the tool tip and tool surface with the AM tool after machining process to represent the tool corrosion. The lower tool pitting corrosion in the AM tool was observed, since the AM tool could produce uniform corrosion and forming fewer pits on the tool surface [32]. Tool surface topology shows that the machining could minimally affect the AM tool the bare tool owing to its ability of lower and tiny craters in surface topology of additive tool [33]. It could be due to a smaller number of manufacturing defects occurred during manufacturing of the tool. The corrosion in bare tool was the pitting corrosion, since the corrosion was not uniform throughout the tool and forming pits on the tool surface. It was observed that the tool surface of the bare tool electrode after the machining process became rougher than the additive tool. The crater size in bare tool was higher than the additive tool. The tool surface and tool tip of additive manufactured was observed with lower corrosion compared to bare tool [34]. It was found that the bare tool could be corroded at a faster rate than that of the additive tool, since the manufacturing of the AM tool electrode could be involved with lower chemical reactance by oxidizing agent in the electrolyte environment to corrode the tool [35]. The corrosion was localized on the areas where machining defects were, which could lead to random distribution of the corrosion. It could lead to lower pitting corrosion with additively manufactured tool electrode in the ECMM process.

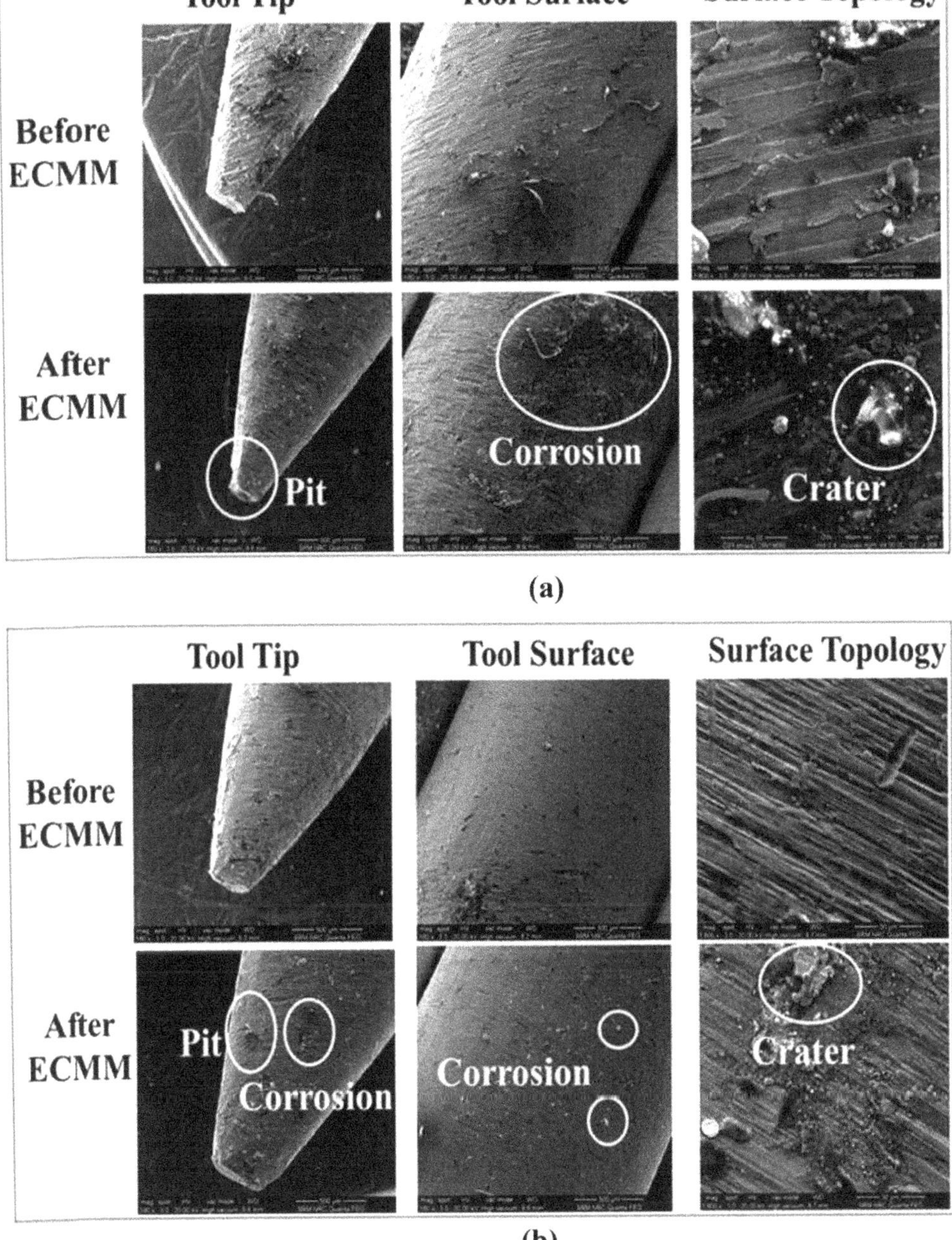

Figure 8. SEM analysis of machined specimens in ECMM (**a**) bare tool (**b**) AM tool.

4. Conclusions

In the present study, an endeavor was performed to find the effects of an additive manufactured tool electrode and process parameters involved in the ECMM process on machining stainless steel for obtaining better performance measures. From the experimental results, the following conclusions were drawn.

- The additive manufactured tool can produce higher *MRR*, since the composition of additive tool has more uniformity with strong atomic bond of metals and higher tool conductivity.
- The additive manufacturing can give considerable dimensional accuracy in terms of circularity and overcut due to increased localization effect and less stray current.
- The lower tool corrosion can be obtained in additively manufactured tool, since the additive tool has porous and less surface defects owing its fabrication of layer-by-layer addition of material.

Author Contributions: Conceptualization, M.T. and R.S.; methodology, M.R. and M.T.; software, M.T.; validation, R.S. and G.T.; formal analysis, M.R. and R.S.; investigation, G.T. and M.R.; resources, R.S.; writing—original draft preparation, M.T. and M.R.; project administration, M.T. and M.R.; funding acquisition, M.R.; All authors have read and agreed to the published version of the manuscript.

Funding: All authors thank to the National Aeronautics and Space Administration (NASA) MUREP Institutional Research Opportunity (MIRO) for providing funding to Navajo Tech Additive Manufacturing Education and Research (NAMER) under NASA Cooperative Agreement 80NSSC19M0227.

Data Availability Statement: Data sharing is not applicable to this article.

Acknowledgments: All authors thank to the National Aeronautics and Space Administration (NASA) MUREP Institutional Research Opportunity (MIRO) for providing funding to Navajo Tech Additive Manufacturing Education and Research (NAMER) under NASA Cooperative Agreement 80NSSC19M0227.

Conflicts of Interest: The authors declare no conflict of interest. The funders had no role in the design of the study; in the collection, analyses, or interpretation of data; in the writing of the manuscript, or in the decision to publish the results.

References

1. Suresh, H.S.; Sudhir, G.B.; Dayanand, S.B. Analysis of electrochemical machining process parameters affecting material removal rate of hastelloy C-276. *Int. J. Adv. Res. Sci. Eng. Technol.* **2014**, *5*, 18–23.
2. Haisch, T.; Mittemeijer, E.; Schultze, J.W. Electrochemical machining of the steel 100Cr6 in aqueous NaCl and NaNO$_3$ solutions: microstructure of surface films formed by carbides. *Electrochim. Acta* **2001**, *47*, 235–241. [CrossRef]
3. Rajurkar, K.P.; Sundaram, M.M.; Malshe, A.P. Review of electrochemical and electro discharge machining. *Procedia CIRP* **2013**, *6*, 13–26. [CrossRef]
4. Sen, M.; Shan, H.S. A review of electrochemical macro- to microhole drilling processes. *Int. J. Mach. Tool. Manuf.* **2005**, *45*, 137–152. [CrossRef]
5. Chiou, Y.C.; Lee, R.T.; Chen, T.J.; Chiou, J.M. Fabrication of high aspect ratio micro-rod using a novel electrochemical micromachining method. *Precis. Eng.* **2012**, *36*, 193–202. [CrossRef]
6. Mileham, A.R.; Jones, R.M.; Harvey, S.J. Changes of valency state during Electrochemical Machining Production Points. *Precis. Eng.* **1982**, *4*, 168–170. [CrossRef]
7. Thanigaivelan, R.; Arunachalam, R.M.; Karthikeyan, B.; Loganathan, P. Electrochemical micromachining of stainless steel with acidified sodium nitrate electrolyte. *Procedia CIRP* **2013**, *6*, 351–355. [CrossRef]
8. De-Silva, A.K.M.; Altena, H.S.J.; McGeough, J.A. Influence of electrolyte concentration on copying accuracy of precision-ECM. *CIRP Ann. Manuf. Technol.* **2013**, *52*, 165–168. [CrossRef]
9. Ayyappan, S.; Sivakumar, K. Experimental investigation on the performance improvement of electrochemical machining process using oxygen-enriched electrolyte. *Int. J. Adv. Manuf. Technol.* **2014**, *75*, 479–487. [CrossRef]
10. Huaiqian, B.; Jiawen, X.; Ying, L. Aviation-oriented Micromachining Technology Micro-ECM in pure Water. *Chin. J. Aeronaut.* **2008**, *218*, 455–461. [CrossRef]
11. Shibuya, N.; Ito, Y.; Natsu, W. Electrochemical Machining of Tungsten Carbide Alloy Micro-pin with NaNO$_3$ Solution. *Int. J. Precis. Eng. Manuf.* **2012**, *13*, 2075–2078. [CrossRef]
12. Tang, L.; Yang, S. Experimental investigation on the electrochemical machining of 00Cr12Ni9Mo4Cu2 material and multi-objective parameters optimization. *Int. J. Adv. Manuf. Technol.* **2013**, *67*, 2909–2916. [CrossRef]

13. Geethapriyan, T.; Muthuramalingam, T.; Moiduddin, K.; Mian, S.M.; Alkhalefah, H.; Umer, U. Performance Analysis of Electrochemical Micro Machining of Titanium (Ti-6Al-4V) Alloy under Different Electrolytes Concentrations. *Materials* **2021**, *11*, 247.
14. Geethapriyan, T.; Kalaichelvan, K.; Muthuramalingam, T. Multi Performance Optimization of Electrochemical Micro-Machining Process Surface Related Parameters on Machining Inconel 718 using Taguchi-grey relational analysis. *Metall. Ital.* **2016**, *4*, 13–19.
15. Geethapriyan, T.; Kalaichelvan, K.; Muthuramalingam, T. Influence of coated tool electrode on drilling Inconel alloy 718 in Electrochemical micro machining. *Procedia CIRP* **2016**, *46*, 127–130. [CrossRef]
16. Muthuramalingam, T.; Akash, R.; Krishnan, S.; Phan, N.H.; Pi, V.N.; Elsheikh, A.H. Surface quality measures analysis and optimization on machining titanium alloy using CO_2 based Laser beam drilling process. *J. Manuf. Process.* **2021**, *62*, 1–6. [CrossRef]
17. Muthuramalingam, T.; Moiduddin, K.; Akash, R.; Krishnan, S.; Mian, S.H.; Ameen, W.; Alkhalefah, H. Influence of process parameters on dimensional accuracy of machined Titanium (Ti-6Al-4V) alloy in Laser Beam Machining Process. *Opt. Laser. Technol.* **2020**, *132*, 106494. [CrossRef]
18. Kirchner, V.; Cagnon, L.; Schuster, R.; Etrl, G. Electrochemical machining of stainless steel microelements with ultrashort voltage pulses. *Appl. Phys. Lett.* **2008**, *79*, 1721–1723. [CrossRef]
19. Spieser, A.; Ivanov, A. Recent developments and research challenges in electrochemical micromachining (μECM). *Int. J. Adv. Manuf. Technol.* **2013**, *69*, 563–581. [CrossRef]
20. Rajurkar, K.P.; Zhu, D.; McGeough, J.A.; Kozak, J.; Silva, A.D. New developments in electro-chemical machining. *CIRP Ann.* **1999**, *48*, 567–579. [CrossRef]
21. Muthuramalingam, T.; Ramamurthy, A.; Sridharan, K.; Ashwin, S. Analysis of surface performance measures on WEDM processed titanium alloy with coated electrodes. *Mater. Res. Express.* **2018**, *5*, 126503. [CrossRef]
22. Rajurkar, K.P.; Zhu, D.; Wei, B. Minimization of machining allowance in electrochemical machining. *CIRP Ann.* **1998**, *47*, 165–168. [CrossRef]
23. Kozak, J.; Rajurkar, K.P.; Makkar, Y. Study of pulse electrochemical micromachining. *J. Manuf. Proc.* **2004**, *6*, 7–14. [CrossRef]
24. Davydov, A.D.; Kabanova, T.B.; Volgin, V.M. Electrochemical machining of titanium. *Review. Russ. J. Electrochem.* **2017**, *53*, 941–965. [CrossRef]
25. Geethapriyan, T.; Muthuramalingam, T.; Kalaichelvan, K. Influence of Process Parameters on Machinability of Inconel 718 by Electrochemical Micromachining Process using TOPSIS Technique. *Arab. J. Sci. Eng.* **2019**, *44*, 7945–7955. [CrossRef]
26. Muthuramalingam, T.; Ramamurthy, A.; Moiduddin, K.; Alkindi, M.; Ramalingam, S.; Alghamdi, O. Enhancing the Surface Quality of Micro Titanium Alloy Specimen in WEDM Process by Adopting TGRA-Based Optimization. *Materials* **2020**, *13*, 1440.
27. Senthilkumar, C.; Ganesan, G.; Karthikeyan, R. Study of electrochemical machining characteristics of Al/SiCp composites. *Int. J. Adv. Manuf. Technol.* **2009**, *43*, 256–263. [CrossRef]
28. Geethapriyan, T.; Kalaichelvan, K.; Jothilingam, A. A Study on Investigating the Effect of various Electrolyte in Electrochemical Machining. *Int. J. Appl. Eng. Res.* **2015**, *10*, 15239–15243.
29. Mingcheng, G.; Yongbin, Z.; Lingchao, M. Electrochemical Micromachining of Square Holes in Stainless Steel in H_2SO_4. *Int. J. Electrochem. Sci.* **2019**, *14*, 414–426. [CrossRef]
30. Saxena, K.K.; Qian, J.; Reynaerts, D. A review on process capabilities of electrochemical micromachining and its hybrid variants. *Int. J. Mach. Tool. Manuf.* **2018**, *127*, 28–56. [CrossRef]
31. Thakur, A.; Tak, M.; Mote, R.G. Electrochemical micromachining behavior on 17-4 PH stainless steel using different electrolytes. *Procedia Manuf.* **2019**, *34*, 355–361. [CrossRef]
32. Leese, R.J.; Ivanov, A. Electrochemical micromachining: An introduction. *Adv. Mech. Eng.* **2016**, *8*, 1–13. [CrossRef]
33. Geethapriyan, T.; Kalaichelvan, K.; Muthuramalingam, T.; Rajadurai, A. Performance Analysis of Process Parameters on Machining α-β Titanium Alloy in Electrochemical Micromachining Process. *Proc. Inst. Mech. E. Part B J. Eng. Manuf.* **2018**, *232*, 1577–1589. [CrossRef]
34. Tang, L.; Li, B.; Yang, S.; Duan, Q.; Kang, B. The effect of electrolyte current density on the electrochemical machining S-03 material. *Int. J. Adv. Manuf. Technol.* **2014**, *71*, 1825–1833. [CrossRef]
35. Chun, K.H.; Kim, S.H.; Lee, E.S. Analysis of the relationship between electrolyte characteristics and electrochemical machinability in PECM on invar (Fe-Ni) fine sheet. *Int. J. Adv. Manuf. Technol.* **2016**, *87*, 3009–3017. [CrossRef]

metals

Article

Microstructure Evaluation, Quantitative Phase Analysis, Strengthening Mechanism and Influence of Hybrid Reinforcements (β-SiCp, Bi and Sb) on the Collective Mechanical Properties of the AZ91 Magnesium Matrix

Song-Jeng Huang, Sikkanthar Diwan Midyeen *, Murugan Subramani and Chao-Ching Chiang

Department of Mechanical Engineering, National Taiwan University of Science and Technology, No. 43, Section 4, Keelung Rd, Daan District, Taipei 10607, Taiwan; sgjghuang@mail.ntust.edu.tw (S.-J.H.); Smurugan2594@gmail.com (M.S.); shoo6667@hotmail.com (C.-C.C.)
* Correspondence: Sikka.ntust@gmail.com; Tel.: +886-966-442-513

Abstract: Gravitational melt-stir casting produced hybrid nano-reinforcements (β-SiCp) and micro-reinforcements (Bi and Sb) of AZ91 composites. SiCp-diffused discontinuous β-$Mg_{17}Al_{12}$ precipitation with a vital factor of SiC was exhibited at the grain boundary region, formulated Mg_3Si throughout the composite and changed the present $Mg_{0.97}Zn_{0.03}$ phases. The creation of Mg_2Si (cubic) and SiC (rhombohedral axes) enhanced the microhardness by 18.60% in a 0.5 wt.% SiCp/AZ91 matrix. The microhardness of 1 wt.% SiCp/AZ91 was slightly reduced after $Mg_{0.97}Zn_{0.03}$ (hexagonal) reduction. The best ultimate tensile value obtained was about 169.33 MPa (increased by 40.10%) in a 0.5 wt.% SiCp/AZ91 matrix. Microelements Bi and Sb developed Mg_3Bi_2, Mg_3Sb_2 and monoclinic C_{60} phases. The best peak yield strength of 82.75 MPa (increased by 19.85%) was obtained with the addition of 0.5 wt.% SiCp/1 wt.% Bi/0.4 wt.% Sb. The mismatch of the coefficient of thermal expansion of segregated particles and the AZ91 matrix, the shear transfer effect and the Orowan effect, combined with the quantitative value of phase existence, improved the compressive strengths of the composites with 0.5 wt.% β-SiCp, 1 wt.% β-SiCp and 0.5 wt.% SiCp/1 wt.% Bi/0.4 wt.% Sb by 2.68%, 6.23% and 8.38%, respectively. Notably, the Charpy impact strengths of 0.5 wt.% and 1 wt.% β-SiCp-added AZ91 composites were enhanced by 236% (2.89 J) and 192% (2.35 J), respectively. The addition of Bi and Sb with SiCp resulted in the formation of a massive phase of brittle Al_6Mn. Al–Mn-based phases (developed huge voids and cavities) remarkably reduced impact values by 80% (0.98 J). The discussion covers the quantitative analyses of X-ray diffraction, optical microscopy and scanning electron microscopy results and fracture surfaces.

Keywords: magnesium alloy; grain and interface; composite; mechanical behaviour; Orowan effect; strengthening mechanism; intermetallic phase; quantitative phase analysis; fractography

Citation: Huang, S.-J.; Diwan Midyeen, S.; Subramani, M.; Chiang, C.-C. Microstructure Evaluation, Quantitative Phase Analysis, Strengthening Mechanism and Influence of Hybrid Reinforcements (β-SiCp, Bi and Sb) on the Collective Mechanical Properties of the AZ91 Magnesium Matrix. *Metals* **2021**, *11*, 898. https://doi.org/10.3390/met11060898

Academic Editor: Claudio Testani

Received: 25 April 2021
Accepted: 27 May 2021
Published: 31 May 2021

1. Introduction

The engineering sector (commercial and military aerospace vehicles, automotive industry and shipbuilding industry) has recently shown broad interest in magnesium alloys [1–4] because of their light weight, which reduces fuel consumption and green gas emission [5]. Magnesium alloys and their composites have low weight (density ρ = 1.74 g/cm³), high specific strength (158 KN m/kg) and superior mechanical strength [6]. Nevertheless, the applications, quality and quantity of magnesium alloys are limited [7] because of the constrained slip arrangement of the hexagonal closed-packed (HCP) crystal lattice [8]; thus, their tensile, compressive and impact properties must be improved. The Federal Aviation Association (FAA) lately reconsidered the practicability of using WE43 and Elektron 21 magnesium alloys because of the increasing demand for novel magnesium alloys. The FAA removed the prohibition against the use of magnesium alloy in seat frame compartments [9].

85

Thus, magnesium alloys with different nano- and microparticles need to be investigated through different mechanical tests.

Several studies have investigated magnesium metal matrix composites (MMCs) to enhance their mechanical properties, such as tensile [10], compression [11] and impact properties [12]. Many factors influence the strength of materials, such as the method of fabrication (ultrasonic stirring method) [13], extrusion of composite and extrusion speed [14], variation in extrusion temperature [13], ageing behaviour [15], variation in ageing temperature [16], annealing behaviour of composites [6] and size of reinforcements [17,18] (microelements and nanoparticles). However, the tensile, compression and impact properties of the hybrid β-SiCp nanoparticle and Bi/Sb microelement reinforcements of AZ91 have not been studied by many. Moreover, the scattering of reinforcements is vital to amend a composite's mechanical strength at a low cost [19,20].

The present paper investigated the hybrid reinforcements of AZ91 (β-SiCp, Bi and Sb) in the as-cast condition by the gravitational melt-stir casting method. The aim of adding SiCp was to settle in the grain boundary and suppress the discontinuous $β$-$Mg_{17}Al_{12}$ precipitation of the AZ91 matrix [21]. In addition, the addition of Bi and Sb formed Mg_3Bi_2 and Mg_3Sb_2 phases, which acted as a straddle in the AZ91 matrix and improved the composite's mechanical properties [22,23]. The mechanical properties of the AZ91 composites with β-SiCp, Bi and Sb were studied by tensile, compression and Charpy impact tests.

The microstructure and phase changes in the composites were studied by X-ray diffraction (XRD), scanning electron microscopy (SEM) and electron-dispersive X-ray spectroscopy (EDS). The qualitative and quantitative phases in the matrix and strengthening mechanisms, such as the CTE, the shear transfer effect, the Orowan effect and fracture surfaces, of all tested samples were studied in detail.

2. Experimental

2.1. Experiment Materials

The present study was conducted on AZ91 magnesium matrix composites reinforced with β-SiCp with different weight percentages (50 nm average particle size), bismuth (50 μm) and antimony (150 μm). The nominal chemical composition of the mother alloy that conforms to AZ91 is shown in Table 1. The same patch production material was used in this study [17,24]. Guangyu Technology Co., Ltd., (Shenzhen, China) and Jiehan Technology Corporation (Taichung) from Taiwan supplied commercial AZ91 alloy and β-SiCp nanoparticles (>99.99% high purity), respectively. Johnson & Annie Co., Ltd., New Taipei, Taiwan, provided bismuth and antimony (>99.99% high-purity) reinforcements.

Table 1. Nominal chemical composition of the investigated AZ91 alloy.

Elements	Al	Zn	Mn	Fe	Si	Cu	Ni	Mg
Wt.% Concentration	8.95	0.84	0.26	0.005	0.009	0.0008	0.0007	Balance

2.2. Composite Preparation

The different composites shown in Table 2 were produced by gravitational melt-stir casting with suitable flux-covered protective atmospheres in a mild steel crucible, as shown in Figure 1.

Table 2. Weight percentages of the investigated composites with reinforcements and their codes.

Composite Code		Wt.% of Reinforcements		
	AZ91	β-SiCp (50 nm)	Bismuth (Bi) (50 μm)	Antimony (Sb) (150 μm)
Code 1	100	0	0	0
Code 2	99.5	0.5	0	0
Code 3	99	1	0	0
Code 4	98.1	0.5	1	0.4

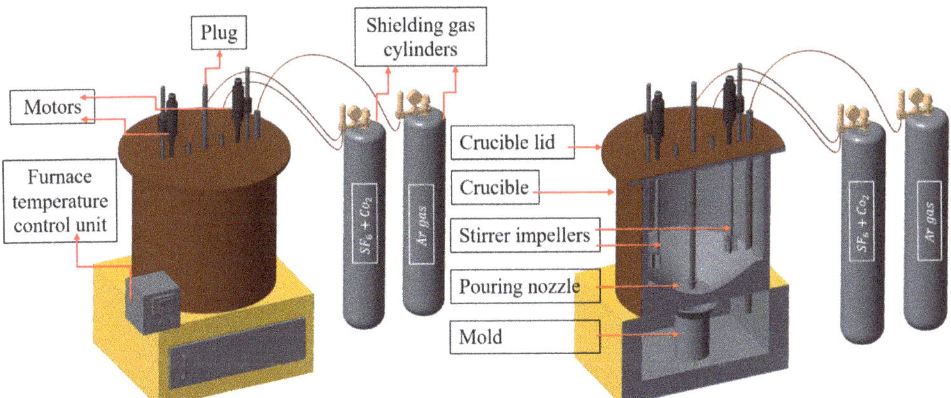

Figure 1. Gravitational melt-stir casting setup and main apparatuses.

Initially, the melts were rapidly heated at 760 °C and the temperature was maintained for 20 min. Then, these melts were stirred for 10 min at 1000 r/min to ensure uniformly dispersed reinforcements in the magnesium composites [12]. After each composite was fully melted, the melts were poured into closed moulds. During the process, the chamber was set up with mixed atmosphere gases of SF_6, CO_2 (1 vol.% and balance) and Ar at 400 °C and 700 °C to terminate oxidation and burning [22].

2.3. Microstructure

According to the American Society for Testing and Materials (ASTM) E3-11 standard of metallographic procedures [25], cubic specimens (10 mm × 10 mm × 10 mm) were sectioned and cut from the centre of the castings using an abrasive cut-off machine. Subsequently, these samples were mechanically ground with silicon carbide polishing papers of various grit sizes (400, 1000, 2000 and 4000 CW) and polished with diamond paste paper (0.3 µm) using water as a lubricant. Before the microstructural investigation, the polished samples were etched for 50 s with an etching solution (75 mL of ethanol, 25 mL of deionized H_2O, 1 mL of nitric and acetic acids). The microstructure and microscopic morphology were analysed by optical microscopy (OM), SEM (JSM-6390L, JEOL, Tokyo, Japan) and EDS. Phase investigation was carried out for all alloy codes by EDS and XRD (Bruker D2 phaser, Billerica, MA, USA) with $CuK\alpha$ radiation at 45 Kv and a current strength of 0.8 mA for composition determination. Measurements were conducted from 20° to 80° at a scan rate of 5°/min with a 12 min scanning span.

2.4. Mechanical Test at Room Temperature

2.4.1. Harness Test

Vickers microhardness was measured in different regions of each specimen using an Akashi MVK-H1 Vickers hardness tester (Mitutoyo, Kawasaki, Japan) with diamond indentation (2.5 mm diameter, 300 gmf applying a load, 10 s dwell time). The microhardness value was calculated based on the average of 10 indentations with two samples.

2.4.2. Tensile Test

The dimensions and location of the dog bone cross-section of the tensile nanocomposite sample are shown in Figure 2. The samples were cut by a water jet cutting machine as per the ASTM E8 standard [26]. Tensile tests were performed on an 810 kN MTS Insight universal testing machine at a constant speed of 0.5 mm/min at room temperature. The alloy composite's mechanical properties (yield strength, ultimate tensile strength (UTS) and elongation) were evaluated by the average of at least three samples to ensure the precision of the final results.

(a) Tensile, compression and impact test samples location & direction.

A-A Section view

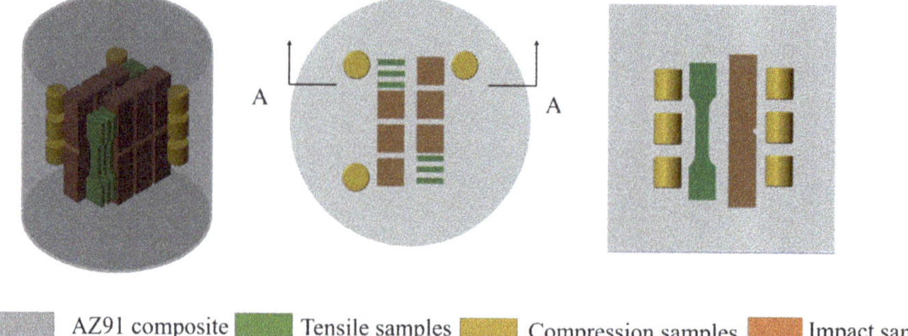

(b) Tensile, compression and impact test samples dimension in mm

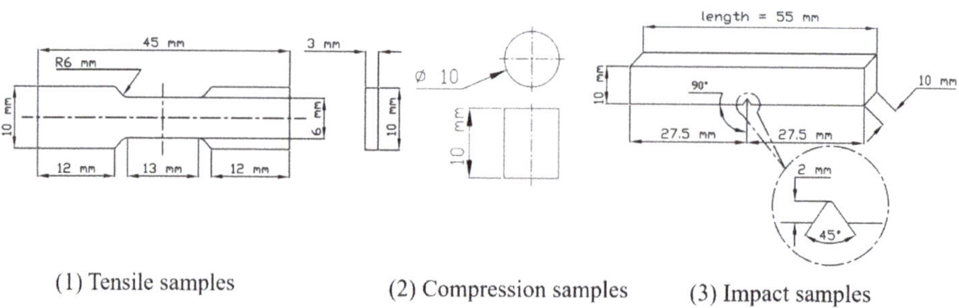

(1) Tensile samples   (2) Compression samples   (3) Impact samples

Figure 2. Schematic illumination of the dimensions of the tensile, compression and impact samples and the direction of the alloy composites.

2.4.3. Compression Test

Three to five cylindrical compression samples (constant length-to-diameter ratio of 1:1) were obtained from the same composites at different locations, as shown in Figure 2, to evaluate the compressive mechanical properties of each composite code as per ASTM E9-19 [27]. The universal testing machine (MTS, 810 kN) (MTS, Eden Prairie, MN, USA) was used at a strain rate of 0.5 mm/min for the compressive tests to evaluate the composites' compressive behaviour at room temperature.

Flow stress (σ) and true strain (ε) values were estimated [17,28] based on the given input load (F), the radius of the cylindrical specimen (R) and the original height of the sample (l_0) in the testing machine according to Equations (1) and (2):

$$\sigma = \frac{F}{\pi R^2} \tag{1}$$

$$\varepsilon = \frac{l_0 - l}{l_0} \tag{2}$$

2.4.4. Impact Test

Charpy impact test specimens (55 mm × 10 mm × 10 mm) were cut and evicted from different regions in each composite by a water jet cutting machine. Moreover, a CNC CHMER electrical discharge machining (CW-43CS, CHMER, Taichung, Taiwan) tool was used to make a V notch, as shown in Figure 2. Three samples were measured to evaluate the precise impact values. The samples were studied with a universal impact tester (model 74) as per ASTM E-23 [29]. All the testing specimens were polished well before testing to terminate uneven alignment, which could lead to improper results. SEM was employed to analyse all fracture surfaces after completing the impact test.

3. Results

3.1. Microhardness

Vickers microhardness values of the hybrid nano- and micro-reinforced Mg MMCs are given in Figure 3 and Table 3. The microhardness of alloy codes 2 and 3 increased by 18.60% and 14.66% compared with the as-cast alloy AZ91, respectively, but decreased with increasing β-SiCp percentage from 0.5 to 1 wt.%. The 3′ microhardness of the AZ91 alloy added with 1 wt.% β-SiCp was slightly reduced; thus, the addition of β-SiCp limited the microhardness of AZ91. The microhardness of alloy code 4 (AZ91 + 0.5 wt.% SiCp + 1 wt.% Bi + 0.4 wt.% Sb) increased by 17.49% compared with that of the pure as-cast AZ91 alloy, as shown in Figure 3 and Table 3. However, the microhardness decreased slightly for alloy code 4 (AZ91 + 0.5 wt.% SiCp + 1 wt.% Bi + 0.4 wt.% Sb) compared with alloy code 2.

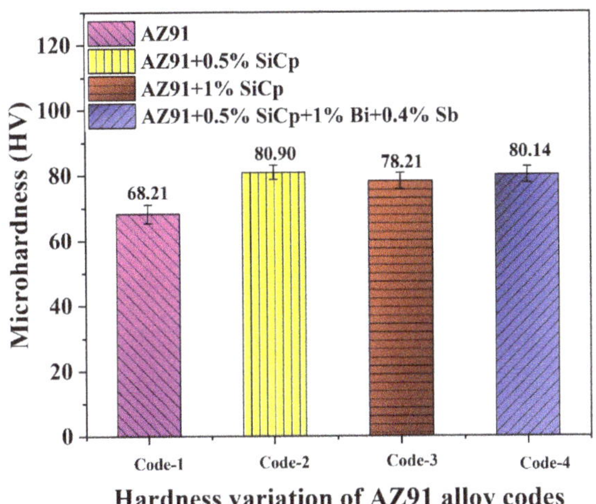

Figure 3. Microhardness of AZ91 alloys with different hybrid nano- and micro-reinforcements.

Table 3. Microhardness variation and increment in microhardness of the alloy codes compared with pure AZ91 alloy.

Alloy Code	Composition	Microhardness (HV)	Increment in Microhardness (%)
Code 1	Pure AZ91	68.21	-
Code 2	AZ91 + 0.5 wt.% SiCp	80.90	18.60
Code 3	AZ91 + 1 wt.% SiCp	78.21	14.66
Code 4	AZ91 + 0.5 wt.% SiCp + 1 wt.% Bi + 0.4 wt.% Sb	80.14	17.49

3.2. Mechanical Properties

3.2.1. Tensile Properties

The room-temperature tensile curves and mechanical tensile properties of pure and hybrid nano-micro-reinforced (β-SiCp, Bi and Sb) as-cast composites are depicted in Figure 4. The comparison of the present study's yield strength, UTS, Young's modulus and elongation with those of other particles reinforced by MMCs in other studies is shown in Table 4. The engineering tensile properties of nanocomposites in as-cast conditions improved remarkably compared with those of the pure AZ91 counterpart.

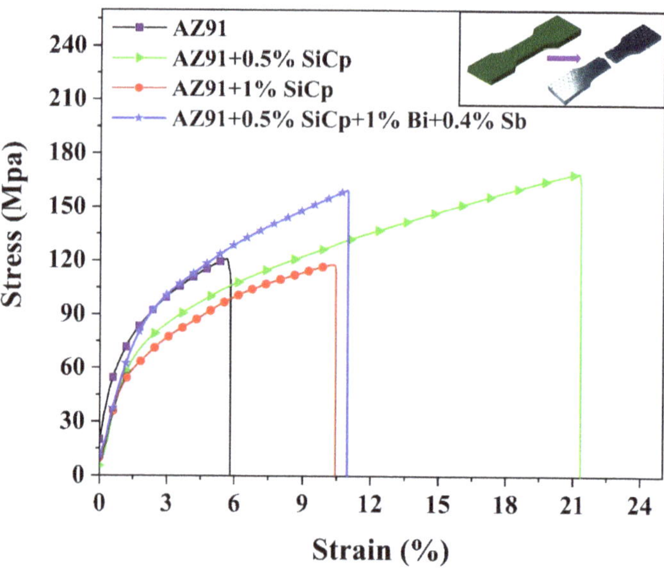

Figure 4. Engineering stress–strain curve for AZ91 alloy composites with different hybrid nano- and micro-reinforcements.

Table 4. Tensile properties of pure AZ91 and AZ91 composites with different hybrid nano- and micro-reinforcements.

Alloy Code	Types of Castings	0.2% Yield Strength (MPa)	UTS (MPa)	Elongation (%)
Code 1	Pure AZ91	69.05 ± 2	120.87 ± 4	6.0 ± 0.1
Code 2	AZ91 + 0.5 wt.% SiCp	72.82 ± 3	169.33 ± 2	21.6 ± 0.3
Code 3	AZ91 + 1 wt.% SiCp	59.38 ± 3	118.07 ± 3	10.6 ± 0.2
Code 4	AZ91 + 0.5 wt.% SiCp + 1 wt.% Bi + 0.4 wt.% Sb	82.75 ± 1	159.60 ± 4	11.2 ± 0.2
Other study [17]	AZ91(as-cast) + 1 wt.% WS$_2$	67.93 ± 3	140.81 ± 6	9.01 ± 0.5
Other study [30]	AZ91(as-cast) +1 wt.% SiCp	116	139	0.78

The variations in the mechanical properties of pure and reinforced AZ91 composites are depicted in Table 4. Tensile strength (169.33 MPa, increased by 40.10%), yield strength (72.82 MPa, increased by 5.47%) and elongation (21.6%, increased by 3.6 times) dramatically increased with an increase in β-SiCp. The increase in elongation reflects that 0.5 wt.% β-SiCp reinforcement increases the formability of the AZ91 matrix alloy substantially. When SiCp content is 0.5 wt.%, the grain refinement effect and the constructively refined Mg17Al12 phase of granulated morphology compounds cause the increase in elongation. Tensile strength (118.07 MPa, decreased by 2.31%) and yield strength (59.38 MPa, decreased by 14.01%) decreased with the addition of 1 wt.% β-SiCp, whereas elongation was preserved (10.6%, increased by 1.77 times). Furthermore, the addition of bismuth and antimony with β-SiCp in alloy code 4 increased the peak yield strength (82.75 MPa, increased by 19.85%) compared with pure AZ91 and the other alloys. In addition, UTS and elongation in alloy code 2 (159.60 MPa, increased by 32.04% compared with pure AZ91) and alloy code 3 (11.2%, increased by 1.87 times) increased compared with pure AZ91. Moreover, the present nanocomposite clearly shows a superior tensile value (UTS, enhanced by 20.25% [17]) compared with other magnesium MMCs [17,30] (such as micro-WS$_2$ and nano-SiCp) or large-sized particles in the as-cast condition, as depicted in Table 4.

3.2.2. Compression Properties

Figure 5 shows the room-temperature engineering compressive stress–strain curves and properties of pure AZ91 and AZ91/β-SiCp/Bi/Sb composites in the as-cast condition. Compressive strength and ductility developed instantaneously with the addition of β-SiCp/Bi/Sb reinforcements, as depicted in Table 5. The compressive properties of alloy codes 2, 3 and 4 improved by 2.68%, 6.23% and 8.38%, respectively, compared with pure AZ91 (345.14 MPa). The present study achieved a higher compressive strength compared with other micro-SiCp-reinforced composites in similar previous studies [31].

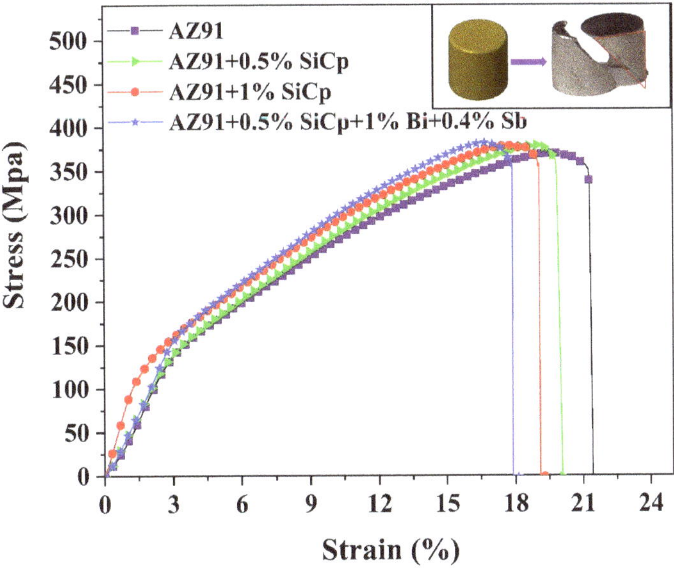

Figure 5. Engineering compressive stress–strain curves of pure AZ91 and AZ91 composites with different hybrid nano- and micro-reinforcements.

Table 5. Compression strength comparison of pure AZ91 and AZ91 composites with different hybrid nano- and micro-reinforcements.

Alloy Code	Types of Castings	Maximum Compressive Strength (MPa)	Compression Ratio (%)
Code 1	Pure AZ91	345.14 ± 13	16.17 ± 0.3
Code 2	AZ91 + 0.5 wt.% SiCp	354.38 ± 10	15.73 ± 0.1
Code 3	AZ91 + 1 wt.% SiCp	366.63 ± 13	15.90 ± 0.3
Code 4	AZ91 + 0.5 wt.% SiCp + 1 wt.% Bi + 0.4 wt.% Sb	374.07 ± 13	15.54 ± 0.3

3.2.3. Impact Properties

The variations in the absorbed Charpy impact energy of pure AZ91 and reinforced (β-SiCp/Bi/Sb) nanocomposites are depicted in Figure 6. Table 6 shows that the absorbed energy expressively improved by 236% (2.89 J) and 192% (2.35 J) after the addition of β-SiCp reinforcements in alloy codes 2 and 3, respectively, compared with pure AZ91 (1.23 J). However, the addition of microelements (Bi and Sb) with the hybrid nanoelement (β-SiCp) in alloy code 4 decreased the impact energy by 80% (0.98 J).

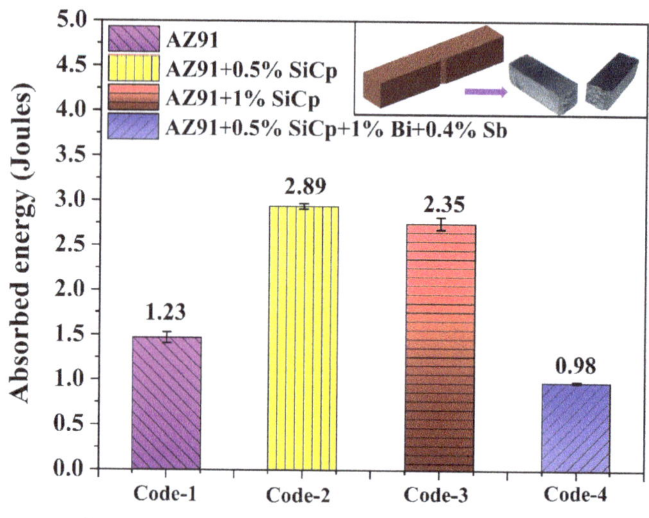

Figure 6. Charpy impact test and absorbed impact energy of pure and modified AZ91 alloys.

Table 6. Variations in the Charpy impact energy of pure AZ91 and AZ91/β-SiCp/Bi/Sb composites.

Alloy Code	Types of Casting	Absorbed Energy (J)	Increment in Absorbed Energy (%)
Code 1	Pure AZ91	1.23	-
Code 2	AZ91 + 0.5 wt.% SiCp	2.89	236
Code 3	AZ91 + 1 wt.% SiCp	2.35	192
Code 4	AZ91 + 0.5 wt.% SiCp + 1 wt.% Bi + 0.4 wt.% Sb	0.98	−80

3.3. X-Ray Diffraction (XRD) Texture Analysis

Figure 7 and Table 7 show the presence of the different composition phases in hybrid nanocomposites (alloy codes 1 to 4) in the as-cast condition, which are confirmed by the XRD test results. The total weight of the phases must be equal to 1 as per Equation (3). The reference intensity ratio (RIR) method was used to estimate the quantitative value of the substantial phases according to Equation (4) [12]. RIRs are the ratios of the most substantial peak of the unknown phase to the peaks of standard phases.

$$\sum_{k=1}^{n} W_k = 1 \tag{3}$$

$$W_\alpha = \frac{I_{i\alpha}}{\mathrm{RIR}_{\alpha c}} \times \left(\sum_{k=1}^{n} \frac{I_{ik}}{\mathrm{RIR}_{kc}} \right)^{1} \times 100\%, \tag{4}$$

where $I_{i\alpha}$ and I_{ik} are the strongest intensities of the *i*-th α and *k* unknown phases, respectively, and $\mathrm{RIR}_{\alpha c}$ and RIR_{kc} are the strongest intensity ratios of the unknown α or *k* phase and corundum *c*.

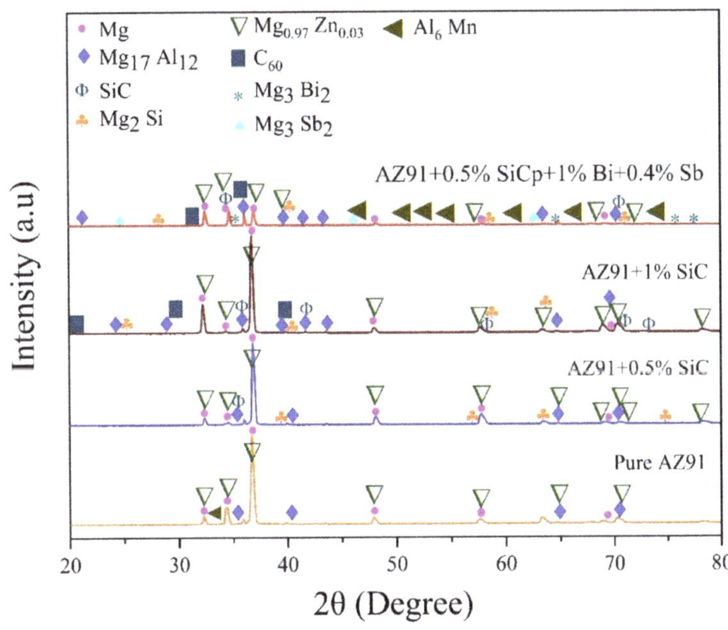

Figure 7. XRD pattern of as-cast AZ91 and AZ91/β-SiCp/Bi/Sb composites.

Table 7. Major intermetallic phases with quantitative values for pure AZ91 and reinforced AZ91 composites.

Alloy Code	Composition	Major Phases	Quantitative Values (%)	Structure
Code 1	Pure AZ91	Mg17Al12	6.81	Cubic
		Mg0.97Zn0.03	91.30	Hexagonal
		Al6Mn	1.89	Orthorhombic
Code 2	AZ91 + 0.5 wt.% SiCp	Mg17Al12	5.18	Cubic
		Mg0.97Zn0.03	90.38	Hexagonal
		SiC	2.52	Rhombo.h.axes
		Mg2Si	1.92	Cubic
Code 3	AZ91 + 1 wt.% SiCp	Mg17Al12	4.64	Cubic
		Mg0.97Zn0.03	86.99	Hexagonal
		SiC	2.69	Rhombo.h.axes
		Mg2Si	2.11	Cubic
		C60	1	Monoclinic
		Al6Mn	2.57	Orthorhombic
Code 4	AZ91 + 0.5 wt.% SiCp + 1 wt.% Bi + 0.4 wt.% Sb	Mg17Al12	15.53	Cubic
		Mg0.97Zn0.03	41.37	Hexagonal
		SiC	14.15	Rhombo.h.axes
		Mg2Si	2.03	Cubic
		C60	1.31	Monoclinic
		Mg3Bi2	15.83	Hexagonal
		Mg3Sb2	4.13	Hexagonal
		Al6Mn	5.64	Orthorhombic

The diffraction peaks of solid solution of α-phase (Mg) and β-phase (Mg$_{17}$Al$_{12}$) were perceived in all the AZ91 matrix composites. The diffraction peaks of SiC, Mg$_2$Si and Mg$_{0.97}$Zn$_{0.03}$ were found in alloy code 2. Additionally, the counts of SiC and Mg$_2$Si were progressively enhanced with the addition of 1 wt.% β-SiCp in alloy code 3. In other words, the formation of Mg$_2$Si substantially dissolved the brittle phase of Mg$_{17}$Al$_{12}$ in alloy code 2 [19,21]. The Mg$_2$Si phase was segregated throughout the composite based on the

XRD pattern [32]. Moreover, the appearance of $Mg_{0.97}Zn_{0.03}$, SiC and Mg_2Si in magnesium composites has also been observed in previous studies [12,30,33–35].

Figure 7 portrays the phases of SiC, Mg_2Si, Mg_3Bi_2, Mg_3Sb_2, C_{60}, $Mg_{0.97}Zn_{0.03}$ and Al_6Mn detected in alloy code 4, and these morphologies of precipitates have been confirmed by earlier studies [12,22,36–38]. A massive amount of Al–Mn-based intermetallic phases and brittle Al_6Mn phases were dispersed throughout the casting. Furthermore, the EDS study confirmed the phases identified in the XRD patterns.

3.4. Microstructural Analysis

The OM and SEM images of the microstructures of pure AZ91 and reinforced AZ91 composites in the as-cast condition are exhibited in Figures 8–11. EDS was used to examine the presence of phases in the MMCs.

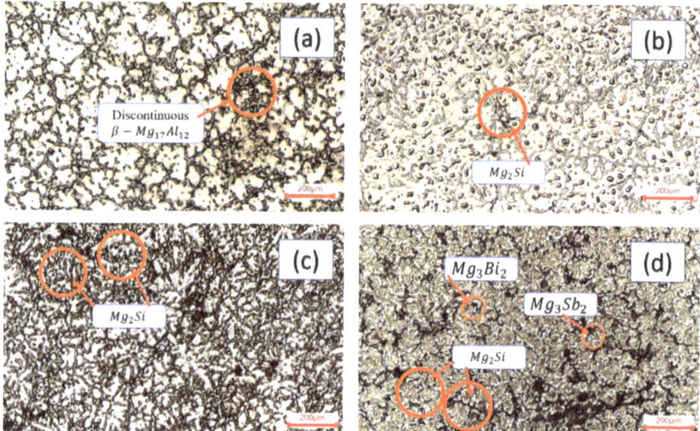

Figure 8. Optical micrographs of (**a**) pure AZ91 matrix, (**b**) AZ91 + 0.5 wt.% SiCp, (**c**) AZ91 + 1 wt.% SiCp and (**d**) AZ91 + 0.5 wt.% SiCp + 1 wt.% Bi + 0.4 wt.% Sb.

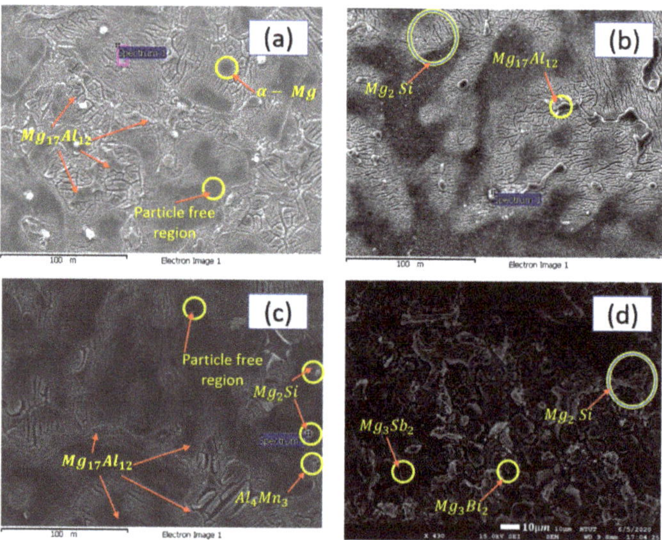

Figure 9. SEM images of (**a**) pure AZ91 matrix, (**b**) AZ91 + 0.5 wt.% SiCp, (**c**) AZ91 + 1 wt.% SiCp and (**d**) AZ91 + 0.5 wt.% SiCp + 1 wt.% Bi + 0.4 wt.% Sb.

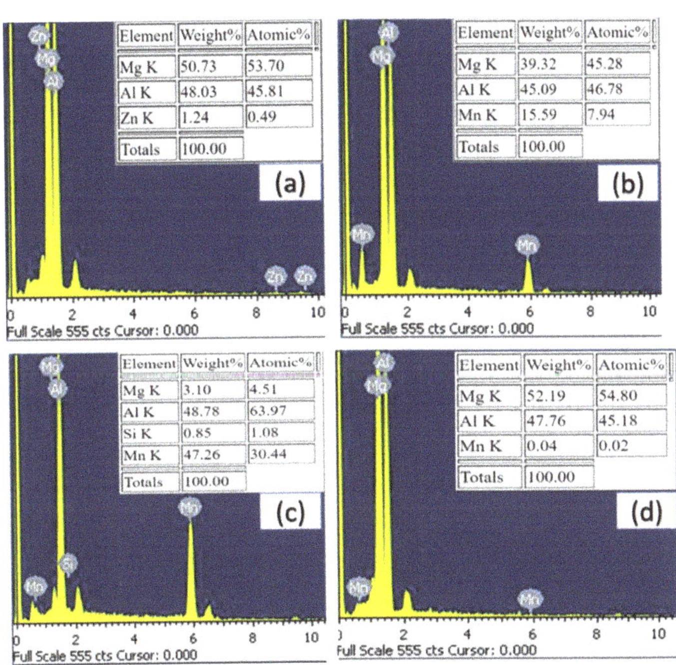

Figure 10. EDS spectra of (**a**) pure AZ9 matrix, (**b**) AZ91 + 0.5 wt.% SiCp and (**c**, **d**) AZ91 + 1 wt.% SiCp.

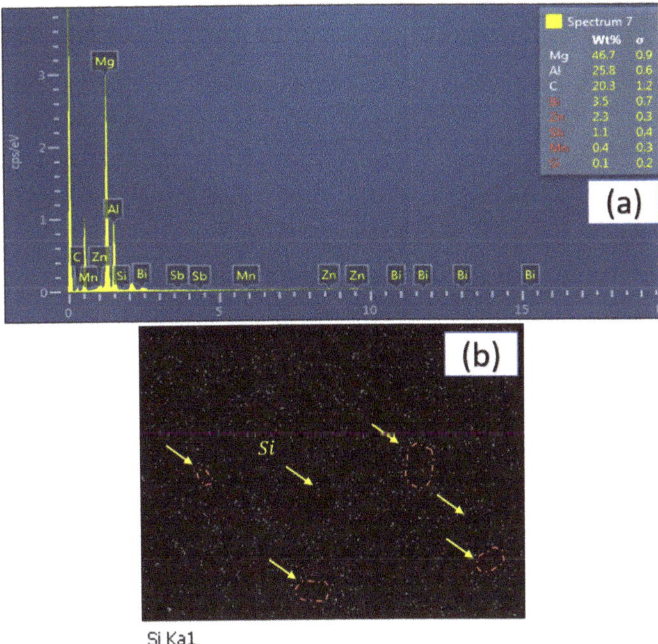

Figure 11. EDS spectra of (**a**) AZ91 + 0.5 wt.% SiCp + 1 wt.% Bi + 0.4 wt.% Sb and (**b**) Si particle dispersion at the boundary region of AZ91 + 0.5 wt.% SiCp.

Alloy code 1 contains an α-Mg matrix and enormous β-Mg17Al12 intermetallic phases. Based on the EDS examination, alloy code 1 contains Mg, Al and a small quantity of Zn. Discontinuous β-Mg17Al12 precipitates arise when eutectic supersaturated Al and Mg cool down and at the eutectic Al supersaturated solution state. New Mg2Si phases developed at the boundary of alloy codes 2 to 4, in addition to the massive β-Mg17Al12 phases. The XRD patterns in Figure 7 indicate the presence of α-Mg, β-Mg17Al12 and Mg2Si. The rough grain structure of alloy code 1 (Figures 8a and 9a) was refined remarkably compared with that of alloy code 2 (Figures 8b and 9b) [39]. On the other hand, the grain sizes of alloy codes 1–4 were calculated by ImageJ software (1.53j, U.S. National Institutes of Health, Bethesda, MD, USA) and the linear intercept method. The grain sizes were estimated from the three images of each alloy code. The average grain sizes of alloy codes 1–4 were recorded as 87, 61, 47 and 81 μm, respectively. The grain refinement increased in alloy codes 2 and 3 remarkably.

In addition, illustrative SEM images (alloy code 2; Figure 11b) show the relatively good scattering of SiC nanoparticles in alloy code 2. The solid solution of discontinuous $Mg_{17}Al_{12}$, which was scattered in a sizeable free region in alloy code 1, as shown in Figure 9a, was dramatically dissolved into alloy code 2, as shown in Figure 9b.

Nevertheless, the addition of 1 wt.% β-SiCp to AZ91 (alloy code 3) can produce many Si nanoparticles and has the potential to reproduce and formulate Mg_2Si (dendritic shape structure of the brittle phase), as shown in Figure 8c. Figure 10c specifies the EDS results for alloy code 3. Alloy code 3 had enormous phases of Al (48.78 wt.%) and Mn (47.26 wt.%), which formed intermediate brittle phases, such as Al_6Mn and others present in the massive $Mg_{17}Al_{12}$ phase in Figure 7 and Table 7. Similarly, Figure 10d shows an enormous solid phase of Mg (52.19 wt.%) and Al (47.76 wt.%). This phase is viewed as a massive discontinuous $Mg_{17}Al_{12}$ phase when the β-SiCp amount is increased to 1 wt.%. Moreover, XRD studies also confirmed the existence of these phases.

The OM and SEM images of alloy code 4 (AZ91 + 0.5 wt.% SiCp + 1 wt.% Bi + 0.4 wt.% Sb) are depicted in Figures 8, 9 and 11. Phases α-Mg, $Mg_{17}Al_{12}$, Mg_2Si, Mg_3Bi_2, Mg_3Sb_2 and C_{60} were identified. The OM and SEM images show that the hexagonal structures of Mg_3Bi_2 and Mg_3Sb_2 are distributed throughout the matrix as rod shapes in the boundary layers in Figures 8d and 9d. The EDS results (Figure 11a) confirmed the presence of Mg (46.7 wt.%), Bi (3.5 wt.%), Sb (1.1 wt.%), Si (0.1 wt.%) and C (20.3 wt.%) elements. Previous studies have also confirmed the creation of Mg_3Bi_2 and Mg_3Sb_2 in the massive β-$Mg_{17}Al_{12}$ phase [22,23,32,35–37,39,40]. The XRD texture for alloy code 4 confirmed these phases, as shown in Figure 7.

4. Discussion

4.1. The Strengthening Mechanism of Microhardness

Research analysis by different research groups [41,42] hasconfirmed that the content of rigid reinforcements in the ductile material matrix enhances the hardness of the composite. The present research also affirmed the same by comparing the microhardness of alloy codes 1 to 4. Each alloy had a different weight percentage and various combinations of reinforcements compared with the as-cast condition of AZ91. For alloy code 2 (0.5 wt.% SiCp added), the microhardness was enhanced by 12.69 HV because of the creation of the brittle texture of Mg_3Bi_2, Mg_2Si (1.92%, cubic) and SiC (2.52%, rhombohedral axes (rhombo.h.axes)). SEM images (Figure 9b,c) reveal that alloy codes 2 and 3 had grain refinement and particle strengthening, which improved the hardness of the composites [43]. Similarly, the microhardness of alloy code 3 (1 wt.% SiCp added) increased by 10 HV compared with alloy code 1. However, the microhardness of alloy code 3 decreased by 2.69 HV, from 90.38% to 86.99%, compared with alloy code 2 because of the formation of $Mg_{0.97}Zn_{0.03}$ (hexagonal) [12]. The values of the quantitative phases were estimated based on the XRD pattern results, as shown in Table 7. Furthermore, the microhardness of alloy code 4 (0.5 wt.% SiCp + 1 wt.% Bi + 0.4 wt.% Sb) increased by 11.93 HV, even with the additional brittle reinforcements Bi and Sb. Here, the microhardness was reduced because of the non-uniform distribution of different reinforcements and their high porosity (Figure 12d) [17]. The presence of reinforcements can be ascribed to the restriction of localized matrix deformation during indentation. The existence of growing agglomeration enhances the porosity concerning the high surface tension and poor wetting properties among the particles and molten melts. The mechanism of the higher viscosity of molten metals and developing propensity of particles to clump increases the agglomeration [44].

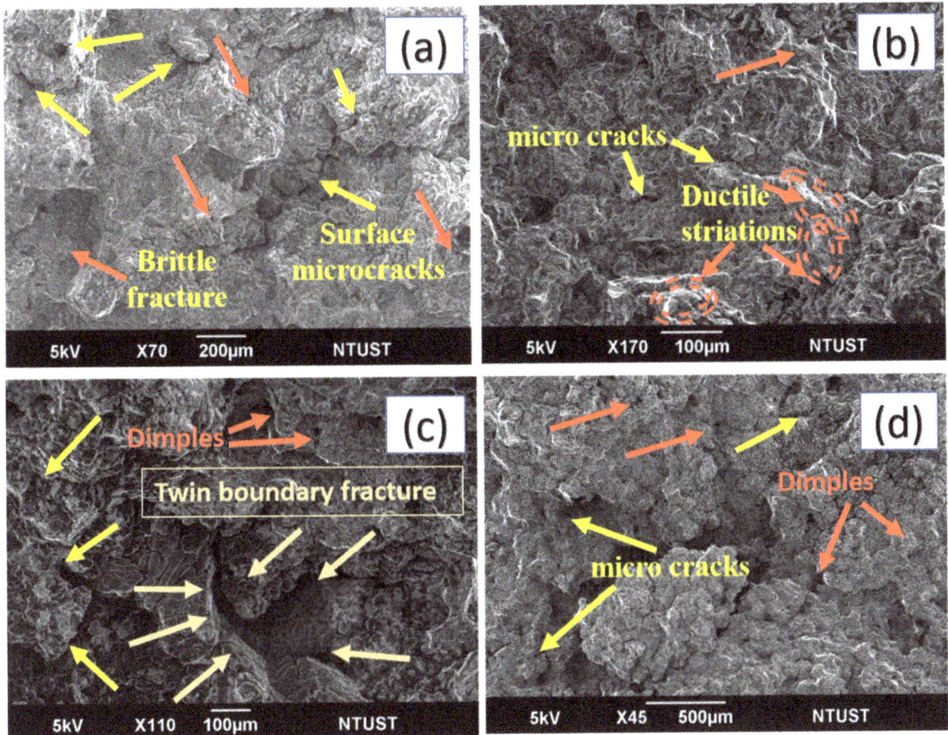

Figure 12. Fractography of the tensile-ruptured surfaces of (**a**) pure AZ91 matrix, (**b**) AZ91 + 0.5 wt.% SiCp, (**c**) AZ91 + 1 wt.% SiCp and (**d**) AZ91 + 0.5 wt.% SiCp + 1 wt.% Bi + 0.4 wt.% Sb.

4.2. Strengthening Mechanism of the Mechanical Properties

As shown in Figures 4–6, the addition of nano-micro-reinforcements to the AZ91 matrix remarkably improved the tensile, compression and impact values compared with the as-cast pure AZ91. The strengthening factors included the mismatch in the CTE, the shear transfer effect of the load, Orowan strengthening development and the quantitative phase present in the composites. These effects contributed to the improvement of mechanical properties, which have also been confirmed by previous studies [17,21,30,33,45,46].

The dislocation effect and induced residual stresses developed because of the CTE mismatch between the AZ91 matrix and reinforcements. The plastic deformation arose in the SiCp vicinity to the Mg alloy because of thermal stresses. The increase in the mechanical properties caused by the CTE can be estimated by Equation (5) [21,45]:

$$\Delta\sigma_{CTE} = \sqrt{3}\beta\mu_m b\sqrt{\frac{12f\Delta\alpha\Delta T}{(1-f)b\bar{d}}},\tag{5}$$

where β is 1.25 (strengthening coefficient), μ_m is 2.64 GPa (elastic modulus of the matrix; μ_m can be calculated by $\mu_m = E_m/[2(1+\nu)]$, where E_m and ν are the Young's modulus and

Poisson's ratio of the matrix AZ91, respectively), $b = 0.32$ nm (Burgers vector) [45], f is 0.0058 (the volume fraction of particles), $\bar{d}$ is the diameter of the particles from Table 2, ΔT is 300 K (test temperature and processing temperature difference) and $\Delta\alpha$ is 25.5×10^{-6} (CTE difference between matrix and reinforcement particles) [47]. Particle size and volume fraction are the critical factors for CTE values, as shown in Equation (5). Thus, $\Delta\sigma_{CTE}$ increased when the particle size of the reinforcements decreased. The influence on strengthening was due to the $\Delta\sigma_{CTE}$ between pure AZ91 and reinforced AZ91, as shown in Table 8.

Table 8. Percentage of strengthening contribution and comparison of the yield strengths of pure AZ91 + 0.5 wt.% SiCp + 1 wt.% Sb + 0.4 wt.% Sb matrix.

Strengthening Mechanism	Values (MPa)	Strengthening Contribution (%)
$\Delta\sigma_{Load}$	0.17	1.712
$\Delta\sigma_{CTE}$	7.67	77.24
$\Delta\sigma_{Orowan}$	2.09	21.05
$\Delta\sigma_{Total}$	9.93	-
$\Delta\sigma_{Yield\ strength\ (AZ91+0.5\ wt.\%\ SiCp+0.1\ wt.\%\ Bi+0.4\ wt.\%Sb)-(Pure\ AZ91)}$	13.70	72.48

Shear load transfers from hard reinforcement particles to the rigid composites during the mechanical testing when sound structural integrity exists between the two phases in the matrix. The load transfer effect is represented by Equation (6) [17,21,33]:

$$\Delta\sigma_I = 0.5 f \sigma_m,\tag{6}$$

where σ_m is the tensile yield strength and $f = 0.00283$ is the volume fraction of the reinforcement particle. The calculated load transfer values are mentioned in Table 8. Load transfer is a critical factor that improves yield strength [21,46]. Increasing load transfer consequently expands the volume of fractions, as shown in Table 8.

As per the Orowan mechanism, finer reinforcements (less than 100 nm) [48] with highly dispersed particles are more practical to enhance the mechanical properties of the composite [49]. The Orowan mechanism developed by the particles is shown in Figure 13. The Orowan mechanism is large and energetically not favourable for the dislocation to cut through in the composites. However, the Orowan mechanism is not vital in micro-sized reinforcements in magnesium MMCs because of the coarse texture and the large spacing of interparticles [48]. As shown in Table 8, the Orowan mechanism contributed about 21.05% against the CTE effect. The collaboration of the non-clipping particles remarkably increased the composite's strength as per the effect of the Orowan strengthening mechanism. The $\Delta\sigma_{Orowan}$ caused by the second-phase dispersion [50] that acts as uniformly distributed spherical reinforcements in the matrix can be estimated by Equations (7) and (8) [30,46,47]:

$$\Delta\sigma_{Orowan} = \frac{0.13 G_m b}{\lambda} \ln\frac{d_p}{2b},\tag{7}$$

$$\lambda = d_p\left[\left(\frac{1}{2f}\right)^{\frac{1}{3}} - 1\right],\tag{8}$$

where $G_m = 2.64$ GPa (shear modulus of the matrix), $b = 0.32$ nm [45,46], $\lambda = 230.56$ nm (the effective planar interparticle spacing), $f = 0.00283$ is the volume fraction of reinforcements calculated from the weight fraction and d_p is the average diameter of the refinement particles from Table 2. The calculated Orowan strengthening mechanism is listed in Table 8.

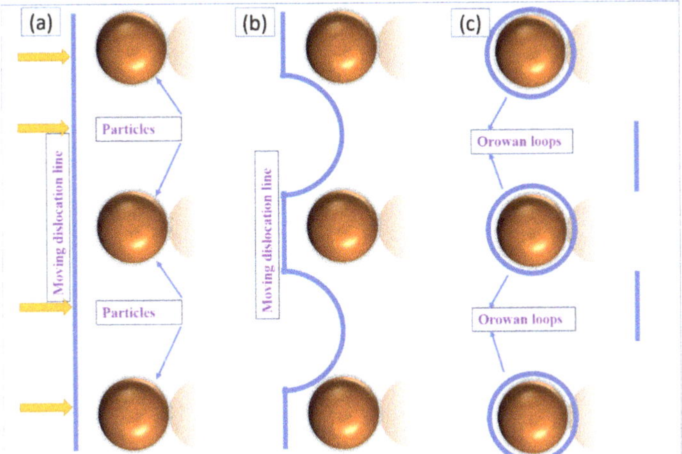

Figure 13. Orowan mechanism of moving the location line from zone (**a**)–(**c**) and Orowan loop formation.

The CTE, the shear transfer effect of the load and Orowan strengthening development in the composites are presented in Table 8. All the effect values (increasing trend of total cumulative) show that the compression test delays crack propagation, dramatically strengthens the interface of the alloys and increases the compression properties (increasing trend), as shown in Table 5 [17].

4.3. Quantitative Phases Influences on Mechanical Properties

An outstanding amount of secondary phases was scattered throughout alloy code 2, and the existence of SiCp increased its microhardness [12]. A refined grain structure means that β-SiCp can restrict grain growth and maintain excellent dispersion (Figure 11b); thus, SiCp is vital for accomplishing excellent mechanical properties, such as tensile and impact properties [33,34].

However, the tensile and impact properties of alloy codes 3 and 4 were reduced in the presence of SiCp, Mg_2Si and C_{60}, which can cause microcracks that coalescence into large cracks [12], as shown in Figure 12c.

Nevertheless, the addition of 1 wt.% β-SiCp (alloy code-3) substantially decreased the mechanical strength (the yield and ultimate properties) because of the sinking and floating comportment of SiCp compared with AZ91 alloy because of its superior density and surface tension. The main prerequisite for these mechanical properties changed because of the presence of new intermetallic Al and Mn phases and massive $Mg_{17}Al_{12}$ (Figure 10c–d) phases. Figure 10c specifies the EDS results for alloy code-3. Alloy code 3 has several combinations of Al (48.78 wt.%) and Mn (47.26 wt.%), which are considered intermediate brittle phases, such as Al_6Mn and others presented in Tables 7 and 9 in the massive Mg_6Al_{12} phase. Similarly, Figure 10d shows an enormous solid phase of Mg (52.19 wt.%) and Al (47.76 wt.%). A massive discontinuous Mg_6Al_{12} phase appeared when the β-SiCp amount increased to 1 wt.%. Moreover, XRD studies also confirmed the existence of these phases, which dominated the mechanical strength of alloy code 3. In addition, previous similar studies have demonstrated that Al–Mn, β-$Mg_{17}Al_{12}$ and Mg_2Si diminish mechanical properties [12,34].

Table 9. Combinations of Al–Mn phases present in alloy codes 1 to 4 based on XRD textures. ([✓] and [×] denote the presence and absence of the respective phase, respectively).

Al–Mn Phases	Alloy Code 1	Alloy Code 2	Alloy Code 3	Alloy Code 4
$Al_{78}Mn_{22}$	✓	×	×	✓
$Al_{0.43}Mn_{0.47}$	✓	×	✓	✓
Al_6Mn [51]	✓	×	×	✓
$Al_{80}Mn_{20}$	×	×	×	✓
$Al_{10}Mn_3$	×	×	×	✓
$Al_{86}Mn_{14}$	×	×	✓	×
$Al_{81}Mn_{19}$	×	✓	✓	✓
$Al_{77}Mn_{23}$	✓	✓	×	×

In the impact test, alloy code 4 had multiple Al–Mn-constructed brittle phases, as shown in Tables 7 and 9, based on the XRD results. The mechanical properties were remarkably reduced because of these Al–Mn-based materials [51]. The results of the tensile test of alloy code 4 displayed in Figure 12d show that these elements can produce substantial amounts of brittle heterogeneous phases of C_{60}, SiCp and Mg_2Si, which can reduce mechanical properties slightly compared to those of alloy code 2 because of the formation of small microcracks (Figure 14d).

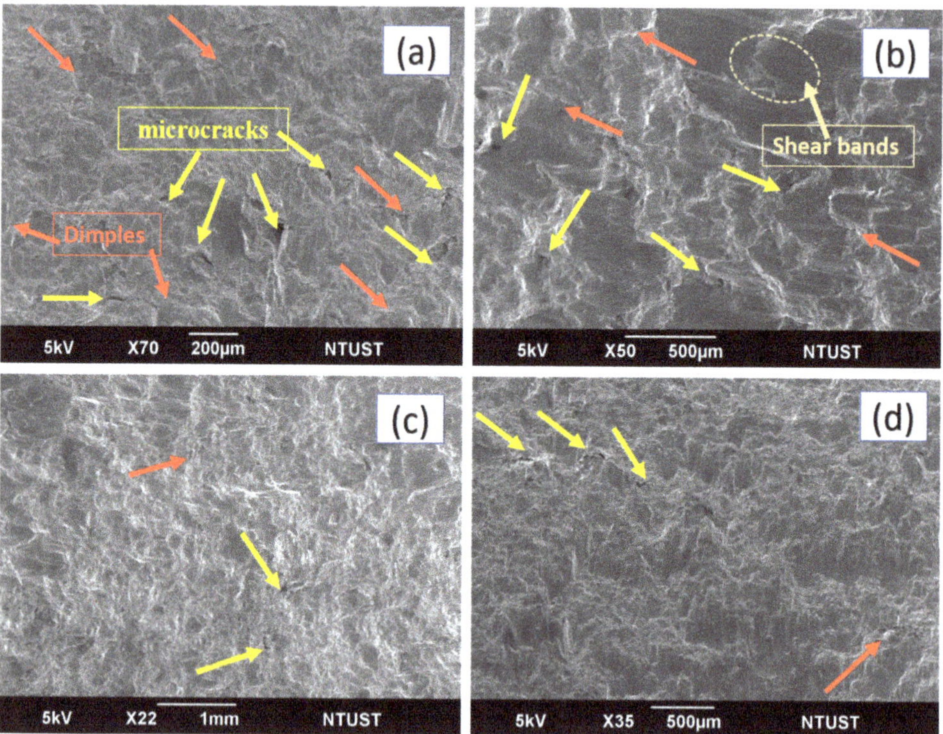

Figure 14. Fractography of the compression-ruptured surfaces of (**a**) pure AZ91 matrix, (**b**) AZ91 + 0.5 wt.% SiCp, (**c**) AZ91 + 1 wt.% SiCp and (**d**) AZ91 + 0.5 wt.% SiCp + 1 wt.% Bi + 0.4 wt.% Sb.

4.4. Fracture Surface Analysis

As depicted in Figures 13–15, the fracture morphology of pure AZ91 and reinforced AZ91 composites (codes 1 to 4) was studied by SEM after tensile, compression and impact tests. The studies reflect the occurrence of a combination of ductile and brittle fractures. β-$Mg_{17}Al_{12}$ obstructed the commencement of cracks and deformation in the AZ91 matrix [52].

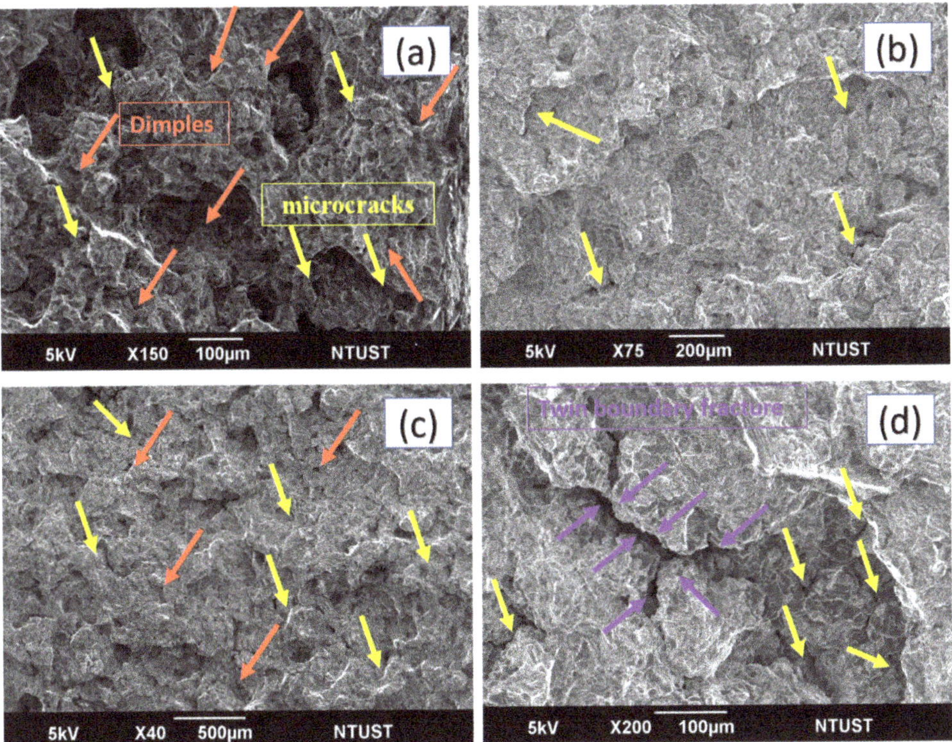

Figure 15. Fractography of the impact-ruptured surfaces of (**a**) pure AZ91 matrix, (**b**) AZ91 + 0.5 wt.% SiCp, (**c**) AZ91 + 1 wt.% SiCp and (**d**) AZ91 + 0.5 wt.% SiCp + 1 wt.% Bi + 0.4 wt.% Sb.

The bunting cracks, microcracks and cavities detected in the 1 wt.% SiCp-reinforcement-added composite alloy (code 3) shown in Figure 12c were more than those in alloy codes 1 and 2 in Figure 12a,b, respectively. Voids are inevitable in the stir casting method with environs but can be governable. These voids exhaust the gas induced in the contiguous particles and augment viscosity. The creation of voids and a secondary phase formulates the stress concentration between particles and matrix. Furthermore, the large number of particles leads to a higher concentration owing to the higher dislocation density [17]. When the load increases during tensile tests, the cleavage facet transfers and breaks the intrinsically brittle structure in the transverse direction of the plane [52]. The load can be moved from the α-Mg matrix to the reinforcement particles; at this moment, an enormous crack is initiated by the reinforcement particles [11]. The tensile values of alloy code 4 decrease slightly because of the reinforcement increments, which can formulate a large number of microcracks. In addition, the presence of Al–Mn-based components (Table 9) can substantially reduce mechanical strength [51].

The number of shear bands (shear tongues) that exist in the fractured surface increased due to the cumulative nano-and micro-reinforcements [11] in alloy codes 2 to 4 compared with alloy code 1, as shown in Figure 14. The cavities and dimples in the fracture surface decreased from alloy codes 2 to 4. In addition, the value of the strengthening mechanism had an increasing trend, as shown in Table 8. The cohesion force between matrix and refinement increased and dimples reduced because of grain refinement. In addition, the interface between plastic flow was due to the rigid reinforcements in the AZ91 matrix [17].

In alloy code 2, β-$Mg_{17}Al_{12}$ diffused along the grain boundary increased the impact property dramatically. The presence of Mg_2Si in alloy code 3 can initiate the formation of micro- and nano-cracks [12]. However, the impact value of alloy code 4 (AZ91 + 0.5 wt.% SiCp + 1 wt.% Bi + 0.4 wt.% Sb) was lower than that of as-cast AZ91 because of the number of reinforcements whose crystal structures existed in the matrix.

The twin boundary fracture, microcracks and surface cracks in the impact test fracture morphology are shown in Figure 15. Alloy code 4 had huge dimples, microcracks and surface cracks compared with the other alloy codes, as shown in Figure 15d. The assortment of C_{60}, SiCp and Mg_2Si heterogeneous phases can reduce the impact strength of the materials, and the presence of huge Al_6Mn phases also plays a vital role in reducing the strength of the impact values [51]. The HCP structures of twin boundary cracks were observed in alloy code 4, and these major boundary cracks, which lead to the reduction of energy values, are shown in Figure 15d.

5. Conclusions

The different amalgamations of the hybrid nanoelement (β-SiCp) and microelements (Bi and Sb) of AZ91 (alloy codes 1 to 4) MMCs were investigated. Microstructure evaluation, phase quantitative analysis, the strengthening mechanism (CTE), the shear transfer effect and the Orowan effect and their impact on mechanical properties (tensile, compression and impact) were studied in detail. The following conclusions are derived from the present study:

- The addition of β-SiCp (0.5 and 1 wt.%) nanoparticles resulted in grain refinement, particle strengthening and creation of quantitative phases Mg_2Si (cubic) and SiC (rhombo.h.axes), which improved the microhardness of the AZ91 matrix by 18.6% and 14.66%, respectively. Similarly, a 17.49% increment in microhardness was generated by adding 0.5% SiCp + 1% Bi + 0.4% Sb to the AZ91 matrix. However, the microhardness of alloy codes 3 and 4 slightly decreased because of the reduction in $Mg_{0.97}Zn_{0.03}$ (hexagonal) brittle phases.
- The small number of 0.5 wt.% β-SiC nanoparticles remarkably improved the tensile and yield values by 40.10% and 5.47% (169.33 and 72.82 MPa, respectively). The best yield strength of 82.75 MPa (increased by 19.85%) was obtained from the AZ91 matrix (alloy code 4) with 0.5 wt.% SiCp, 1 wt.% Bi and 0.4 wt.% Sb as reinforcements. Strength was related to the mismatch of the CTE and the existence of quantitative phases of Mg_2Si, Mg_3Bi_2, Mg_3Sb_2 and $Mg_{0.97}Zn_{0.03}$ in the AZ91 matrix.
- Compression properties increased by 2.68%, 6.23% and 8.38% for alloy codes 2 to 4, respectively. CTE augmentation; the shear transfer effect of the load; the Orowan strengthening effect and the dispersion of quantitative phases Mg_2Si, Mg_3Bi and Mg_3Sb_2 in the composites presented delays in crack propagation and occurrence at the particle–matrix interface.
- The addition of β-SiC remarkably reduced brittle Al–Mn-based phases; thus, the absorbed Charpy impact energy improved by 236% (2.89 J) and 192% (2.35 J) for alloy codes 2 and 4, respectively.
- Alloy code 4 (0.5 wt.% SiCp + 1 wt.% Bi + 0.4 wt.% Sb) gained impact energy negatively, 0.98 J (-80%), because of the increment in reinforcement quantity and the creation of C_{60}, SiCp and Mg_2Si phases. Al–Mn-based brittle phases, such as Al_6Mn, were observed in the XRD pattern. A large number of dimples and microcracks were created in alloy code 4's impact test fracture morphology because of the sinking and floating comportments of SiCp, Sb and Bi.

Particle debonding, the coalescence of voids and the nucleation growth of ductile and brittle fractures were detected in tensile morphology. Debonding of microcracks was perceived in the impact of the fragile fracture surface.

Author Contributions: Conceptualization, S.-J.H. and S.D.M.; methodology, S.-J.H. and S.D.M.; software, S.D.M., M.S.; validation, S.-J.H., S.D.M. and M.S.; formal analysis, S.D.M.; resources, C.-C.C.; data curation, S.D.M.; writing—original draft preparation, S.D.M.; writing—review and editing, S.D.M.; visualization, S.D.M.; supervision, S.-J.H.; project administration, S.D.M. and C.-C.C.; funding acquisition, S.-J.H. All authors have read and agreed to the published version of the manuscript.

Funding: This research was funded by Ministry of Science and Technology of the Republic of China (project no. MOST-108-2218-E-011-011-009).

Data Availability Statement: Not applicable.

Conflicts of Interest: The authors declare no conflict of interest.

References

1. Yang, Z.R.; Wang, S.Q.; Gao, M.J.; Zhao, Y.T.; Chen, K.M.; Cui, X.H. A new-developed magnesium matrix composite by reactive sintering. *Compos. Part A Appl. Sci. Manuf.* **2008**, *39*, 1427–1432. [CrossRef]
2. Meng, F.; Lv, S.; Yang, Q.; Qin, P.; Zhang, J.; Guan, K.; Huang, Y.; Hort, N.; Li, B.; Liu, X.; et al. Developing a die casting magnesium alloy with excellent mechanical performance by controlling intermetallic phase. *J. Alloys Compd.* **2019**, *795*, 436–445. [CrossRef]
3. Joost, W.J.; Krajewski, P.E. Towards magnesium alloys for high-volume automotive applications. *Scr. Mater.* **2017**, *128*, 107–112. [CrossRef]
4. Jin, T.; Zhou, Z.; Qiu, J.; Wang, Z.; Zhao, D.; Shu, X.; Yan, S. Investigation on the yield behavior of AZ91 magnesium alloy. *J. Alloys Compd.* **2018**, *738*, 79–88. [CrossRef]
5. Lin, X.Z.; Chen, D.L. Strain controlled cyclic deformation behavior of an extruded magnesium alloy. *Mater. Sci. Eng. A* **2008**, *496*, 106–113. [CrossRef]
6. Lee, S.W.; Park, S.H. Static recrystallization mechanism in cold-rolled magnesium alloy with off-basal texture based on quasi in situ EBSD observations. *J. Alloys Compd.* **2020**, *844*, 156185. [CrossRef]
7. Martynenko, N.; Lukyanova, E.; Serebryany, V.; Prosvirnin, D.; Terentiev, V.; Raab, G.; Dobatkin, S.; Estrin, Y. Effect of equal channel angular pressing on structure, texture, mechanical and in-service properties of a biodegradable magnesium alloy. *Mater. Lett.* **2019**, *238*, 218–221. [CrossRef]
8. Mironov, S.; Onuma, T.; Sato, Y.S.; Kokawa, H. Microstructure evolution during friction-stir welding of AZ31 magnesium alloy. *Acta Mater.* **2015**, *100*, 301–312. [CrossRef]
9. Bian, M.; Huang, X.; Chino, Y. Improving flame resistance and mechanical properties of magnesium–silver–calcium sheet alloys by optimization of calcium content. *J. Alloys Compd.* **2020**, *837*, 155551. [CrossRef]
10. Wang, H.Y.; Rong, J.; Yu, Z.Y.; Zha, M.; Wang, C.; Yang, Z.Z.; Bu, R.Y.; Jiang, Q.C. Tensile properties, texture evolutions and deformation anisotropy of as-extruded Mg-6Zn-1Zr magnesium alloy at room and elevated temperatures. *Mater. Sci. Eng. A* **2017**, *697*, 149–157. [CrossRef]
11. Yang, C.; Zhang, B.; Zhao, D.; Wang, X.; Sun, Y.; Xu, X.; Liu, F. Microstructure evolution of as-cast AlN/AZ91 composites and room temperature compressive properties. *J. Alloys Compd.* **2019**, *774*, 573–580. [CrossRef]
12. Huang, S.J.; Ali, A.N. Effects of heat treatment on the microstructure and microplastic deformation behavior of SiC particles reinforced AZ61 magnesium metal matrix composite. *Mater. Sci. Eng. A* **2018**, *711*, 670–682. [CrossRef]
13. Nie, K.; Guo, Y.; Deng, K.; Kang, X. High strength TiCp/Mg-Zn-Ca magnesium matrix nanocomposites with improved formability at low temperature. *J. Alloys Compd.* **2019**, *792*, 267–278. [CrossRef]
14. Sun, X.F.; Wang, C.J.; Deng, K.K.; Nie, K.B.; Zhang, X.C.; Xiao, X.Y. High strength SiCp/AZ91 composite assisted by dynamic precipitated Mg17Al12 phase. *J. Alloys Compd.* **2018**, *732*, 328–335. [CrossRef]
15. Xiao, P.; Gao, Y.; Yang, X.; Xu, F.; Yang, C.; Li, B.; Li, Y.; Liu, Z.; Zheng, Q. Processing, microstructure and ageing behavior of in-situ submicron TiB2 particles reinforced AZ91 Mg matrix composites. *J. Alloys Compd.* **2018**, *764*, 96–106. [CrossRef]
16. Wang, X.J.; Hu, X.S.; Liu, W.Q.; Du, J.F.; Wu, K.; Huang, Y.D.; Zheng, M.Y. Ageing behavior of as-cast SiCp/AZ91 Mg matrix composites. *Mater. Sci. Eng. A* **2017**, *682*, 491–500. [CrossRef]
17. Huang, S.J.; Abbas, A. Effects of tungsten disulfide on microstructure and mechanical properties of AZ91 magnesium alloy manufactured by stir casting. *J. Alloys Compd.* **2020**, *817*, 153321. [CrossRef]
18. Xiao, P.; Gao, Y.; Xu, F.; Yang, C.; Li, Y.; Liu, Z.; Zheng, Q. Tribological behavior of in-situ nanosized TiB2 particles reinforced AZ91 matrix composite. *Tribol. Int.* **2018**, *128*, 130–139. [CrossRef]
19. Ganguly, S.; Mondal, A.K. Influence of SiC nanoparticles addition on microstructure and creep behavior of squeeze-cast AZ91-Ca-Sb magnesium alloy. *Mater. Sci. Eng. A* **2018**, *718*, 377–389. [CrossRef]

20. Zhu, Y.P.; Jin, P.P.; Fei, W.D.; Xu, S.C.; Wang, J.H. Effects of Mg2B2O5 whiskers on microstructure and mechanical properties of AZ31B magnesium matrix composites. *Mater. Sci. Eng. A* **2017**, *684*, 205–212. [CrossRef]
21. Sun, X.F.; Wang, C.J.; Deng, K.K.; Kang, J.W.; Bai, Y.; Nie, K.B.; Shang, S.J. Aging behavior of AZ91 matrix influenced by 5 μm SiCp: Investigation on the microstructure and mechanical properties. *J. Alloys Compd.* **2017**, *727*, 1263–1272. [CrossRef]
22. Elen, L.; Zengin, H.; Turen, Y.; Turan, M.E.; Sun, Y.; Ahlatci, H. Effects of bismuth (bi) additions on microstructure and mechanical properties of AZ91 alloy. In Proceedings of the Metal 2015—24th International Conference on Metallurgy and Materials, Conference Proceedings, Brno, Czech Republic, 3–5 June 2015; pp. 1124–1128.
23. Guangyin, Y.; Yangshan, S.; Wenjiang, D. Effect of bismuth and antimony additions on the microstructure and mechanical properties of AZ91 magnesium alloy. *Mater. Sci. Eng. A* **2001**, *308*, 38–44. [CrossRef]
24. Abbas, A.; Huang, S.J. Investigation of severe plastic deformation effects on microstructure and mechanical properties of WS2/AZ91 magnesium metal matrix composites. *Mater. Sci. Eng. A* **2020**, *780*, 139211. [CrossRef]
25. ASTM International. E3 Preparation of Metallographic Specimens. *Annu. B ASTM Stand.* **2017**, *11*, 1–17.
26. ASTM E8 ASTM E8/E8M standard test methods for tension testing of metallic materials 1. *Annu. B. ASTM Stand. 4* **2010**, 1–27. [CrossRef]
27. E9-19 Standard Test Methods of Compression Testing of Metallic Materials at Room Temperature. *ASTM B. Stand.* **2019**, 1–10. [CrossRef]
28. Yi, Y.S.; Li, R.F.; Xie, Z.Y.; Meng, Y.; Sugiyama, S. Effects of reheating temperature and isothermal holding time on the morphology and thixo-formability of SiC particles reinforced AZ91 magnesium matrix composite. *Vacuum* **2018**, *154*, 177–185. [CrossRef]
29. ASTM International. ASTM E23—18, Standard Test Methods for Notched Bar Impact Testing of Metallic Materials. *ASTM Int.* **2018**, 1–26. [CrossRef]
30. Qiao, X.G.; Ying, T.; Zheng, M.Y.; Wei, E.D.; Wu, K.; Hu, X.S.; Gan, W.M.; Brokmeier, H.G.; Golovin, I.S. Microstructure evolution and mechanical properties of nano-SiCp/AZ91 composite processed by extrusion and equal channel angular pressing (ECAP). *Mater. Charact.* **2016**, *121*, 222–230. [CrossRef]
31. Viswanath, A.; Dieringa, H.; Ajith Kumar, K.K.; Pillai, U.T.S.; Pai, B.C. Investigation on mechanical properties and creep behavior of stir cast AZ91-SiCp composites. *J. Magnesuim Alloy.* **2015**, *3*, 16–22. [CrossRef]
32. Srinivasan, A.; Swaminathan, J.; Pillai, U.T.S.; Guguloth, K.; Pai, B.C. Effect of combined addition of Si and Sb on the microstructure and creep properties of AZ91 magnesium alloy. *Mater. Sci. Eng. A* **2008**, *485*, 86–91. [CrossRef]
33. Zhou, X.; Su, D.; Wu, C.; Liu, L. Tensile mechanical properties and strengthening mechanism of hybrid carbon nanotube and silicon carbide nanoparticle-reinforced magnesium alloy composites. *J. Nanomater.* **2012**, *2012*, 851862. [CrossRef]
34. Zhao, W.; Huang, S.J.; Wu, Y.J.; Kang, C.W. Particle size and particle percentage effect of AZ61/SiCp magnesium matrix micro- and nano-composites on their mechanical properties due to extrusion and subsequent annealing. *Metals* **2017**, *7*, 293. [CrossRef]
35. Srinivasan, A.; Pillai, U.T.S.; Pai, B.C. Effects of elemental additions (Si and Sb) on the ageing behavior of AZ91 magnesium alloy. *Mater. Sci. Eng. A* **2010**, *527*, 6543–6550. [CrossRef]
36. Guangyin, Y.; Yangshan, S.; Weiming, Z. Improvements of tensile strength and creep resistance of Mg-9Al based alloy with antimony addition. *J. Mater. Sci. Lett.* **1999**, *18*, 2055–2057. [CrossRef]
37. Wang, Q.; Chen, W.; Ding, W.; Zhu, Y.; Mabuchi, M. Effect of Sb on the Microstructure and Mechanical Properties of AZ91 Magnesium Alloy. *Metall. Mater. Trans. A* **2001**, *32*, 787–794. [CrossRef]
38. Bag, A.; Zhou, W. Tensile and fatigue behavior of AZ91D magnesium alloy. *J. Mater. Sci. Lett.* **2001**, *20*, 457–459. [CrossRef]
39. Srinivasan, A.; Swaminathan, J.; Gunjan, M.K.; Pillai, U.T.S.; Pai, B.C. Effect of intermetallic phases on the creep behavior of AZ91 magnesium alloy. *Mater. Sci. Eng. A* **2010**, *527*, 1395–1403. [CrossRef]
40. Srinivasan, A.; Pillai, U.T.S.; Pai, B.C. Microstructure and mechanical properties of Si and Sb added AZ91 magnesium alloy. *Metall. Mater. Trans. A Phys. Metall. Mater. Sci.* **2005**, *36*, 2235–2243. [CrossRef]
41. Nguyen, Q.B.; Gupta, M. Increasing significantly the failure strain and work of fracture of solidification processed AZ31B using nano-Al2O3 particulates. *J. Alloys Compd.* **2008**, *459*, 244–250. [CrossRef]
42. Inegbenebor, A.O.; Bolu, C.A.; Babalola, P.O.; Inegbenebor, A.I.; Fayomi, O.S.I. Aluminum Silicon Carbide Particulate Metal Matrix Composite Development Via Stir Casting Processing. *Silicon* **2018**, *10*, 343–347. [CrossRef]
43. Xiao, P.; Gao, Y.; Xu, F.; Yang, S.; Li, B.; Li, Y.; Huang, Z.; Zheng, Q. An investigation on grain refinement mechanism of TiB2 particulate reinforced AZ91 composites and its effect on mechanical properties. *J. Alloys Compd.* **2019**, *780*, 237–244. [CrossRef]
44. Aydin, F.; Sun, Y.; Emre Turan, M. The Effect of TiB2 Content on Wear and Mechanical Behavior of AZ91 Magnesium Matrix Composites Produced by Powder Metallurgy. *Powder Metall. Met. Ceram.* **2019**, *57*, 564–572. [CrossRef]
45. Deng, K.; Shi, J.; Wang, C.; Wang, X.; Wu, Y.; Nie, K.; Wu, K. Microstructure and strengthening mechanism of bimodal size particle reinforced magnesium matrix composite. *Compos. Part A Appl. Sci. Manuf.* **2012**, *43*, 1280–1284. [CrossRef]
46. Habibnejad-Korayem, M.; Mahmudi, R.; Poole, W.J. Enhanced properties of Mg-based nano-composites reinforced with Al2O3 nano-particles. *Mater. Sci. Eng. A* **2009**, *519*, 198–203. [CrossRef]
47. Huang, S.J.; Lin, P.C. Grain refinement of AM60/Al2O3p magnesium metal-matrix composites processed by ECAE. *Kov. Mater.* **2013**, *51*, 357–366.
48. Karbalaei Akbari, M.; Baharvandi, H.R.; Shirvanimoghaddam, K. Tensile and fracture behavior of nano/micro TiB2 particle reinforced casting A356 aluminum alloy composites. *Mater. Des.* **2015**, *66*, 150–161. [CrossRef]

49. Huang, S.J.; Ho, C.H.; Feldman, Y.; Tenne, R. Advanced AZ31 Mg alloy composites reinforced by WS2 nanotubes. *J. Alloys Compd.* **2016**, *654*, 15–22. [CrossRef]
50. Saba, F.; Zhang, F.; Liu, S.; Liu, T. Reinforcement size dependence of mechanical properties and strengthening mechanisms in diamond reinforced titanium metal matrix composites. *Compos. Part B Eng.* **2019**, *167*, 7–19. [CrossRef]
51. Tokarski, T. Effect of rapid solidification on the structure and mechanical properties of AZ91 magnesium alloy. *Solid State Phenom.* **2012**, *186*, 120–123. [CrossRef]
52. Abbas, A.; Huang, S.J. ECAP effects on microstructure and mechanical behavior of annealed WS2/AZ91 metal matrix composite. *J. Alloys Compd.* **2020**, *835*, 155466. [CrossRef]

Article

Experimental Investigation of Material Properties in FSW Dissimilar Aluminum-Steel Lap Joints

Michelangelo Mortello [1,*], Matteo Pedemonte [1], Nicola Contuzzi [2] and Giuseppe Casalino [2]

[1]   Istituto Italiano della Saldatura, Lungobisagno Istria 15, 16141 Genova, Italy; matteo.pedemonte@iis.it

[2]   Dipartimento di Meccanica, Matematica e Management, Polytechnic University of Bari, Via Orabona 4, 70125 Bari, Italy; nicola.contuzzi@poliba.it (N.C.); giuseppe.casalino@poliba.it (G.C.)

*   Correspondence: Michelangelo.mortello@iis.it; Tel.: +39-010-8341-539

Abstract: The friction stir lap welding of AA5083 H111 aluminum alloy and S355J2 grade DH36 structural steel was investigated. A polycrystalline cubic boron nitride with tungsten and rhenium additives tool was used. According to visual inspection, radiographic examination, and tensile test, it was observed that the best results were obtained for rotation speeds of about 700–800 rpm, with a feed speed ranging between 1.3 and 1.9 mm/s. From the fatigue tests, it is possible to state that there was a preferential propagation of cracks in the part of the aluminum alloy base material. Furthermore, a different response to fatigue stress for samples extracted from the same weld at different positions was observed, which introduces an overall variability in weld behavior along the welding direction. The specimens obtained in the second part of the weld endured a larger number of cycles before reaching failure, which can be related to progressively varying thermal conditions, dissipation behavior, and better metal coupling as the tool travels along the welding line.

Keywords: friction stir welding; mechanical properties; dissimilar aluminum-steel lap joints

Citation: Mortello, M.; Pedemonte, M.; Contuzzi, N.; Casalino, G. Experimental Investigation of Material Properties in FSW Dissimilar Aluminum-Steel Lap Joints. *Metals* **2021**, *11*, 1474. https://doi.org/10.3390/met11091474

Academic Editors: Evgeny A. Kolubaev and António Bastos Pereira

Received: 13 July 2021
Accepted: 10 September 2021
Published: 16 September 2021

Publisher's Note: MDPI stays neutral with regard to jurisdictional claims in published maps and institutional affiliations.

1. Introduction

In recent years, aluminum has experienced a high demand from industries because of its outstanding properties, including high resistance-to-weight ratio, ductility, and high corrosion resistance, which allowed it to be exploited for shipbuilding, automotive applications, and aeronautics. Because of their high strength and lightweight properties, magnesium-based aluminum alloys enable the production of highly performing and extremely durable hulls, decks, and bulkheads. In particular, Aluminum 5083 is known for exceptional high-resistance performance in harsh environments, including seawater and industrial-chemical environments. It also retains exceptional strength after welding. Typical applications are shipbuilding, rail cars, vehicle bodies, tip truck bodies, mine skips, and cages and pressure vessels.

Despite the mentioned benefits and the large use for different assemblies, fusion welding of aluminum alloys is challenging since it can lead to metallurgical defects, mechanical discontinuities, and reductions in mechanical properties. Solid-state welding introduces the key advantage of avoiding the fusion and limiting thermal effect on material properties, leading to improved weld quality.

Within naval sector applications, the use of steel is also well recognized because of several advantages including weight reduction of about 40% compared with a similar wooden structure, simplicity of structuring, larger load capacity, larger impermeability of the planking and absence of caulking, higher durability, and ease of repair [1]. For example, S355J2 grade DH36 structural steel is used in the naval sector as a structural steel for the hull; it possesses good characteristics of toughness, machinability, and weldability, together with a good impact resistance (also at sub-zero temperatures).

Joining aluminum to steel enables the ability to design and fabricate components whose properties are customized to locally varying environmental conditions. However,

107

dissimilar joining is highly affected by differences in thermal properties, limited mutual solubility, and metallurgical compatibility. These challenges have raised the importance of assessing the use of solid-state approaches, like Friction Stir Welding (FSW) for aluminum-to steel assemblies.

Friction Stir Welding (FSW) was invented and patented in 1991 [2], and in the early stages of development companies have implemented the process predominantly in the fabrication of aluminum components and panels [3].

FSW processes for steels require improvements to tool material technology and process control. The process economy for steel welding has not been fully established; moreover, the robustness of the process for shipyard applications requires further consideration.

It is necessary to consider that the selection of the material to be adopted for the construction of the tool is limited due to the high temperatures reached during the process (about 1000–1200 °C). An essential requirement for successfully carrying out the welding is that the tool does not suffer degradation actions caused by wear, deformations, microstructural instability, fracture, or reactivity with the material that constitutes the workpiece. The tool is considered as the most critical component for performing a highly efficient defect-free joint [4]. During FSW of Al to steel, Zhou et al. [5] totally plunged the pin into the Al plate in order to avoid tool wear and this produced a diffusion layer at the interface, rather than an intermetallic layer with enhanced atomic migration. Refractory alloys based on the tungsten-rhenium system [6] are widely adopted for steel processing; they consist of a ceramic solution based on polycrystalline cubic boron nitride (PCBN) [7] and Co-based alloy with the γ/γ' microstructure strengthened by precipitating intermetallics, $Co_3(Al, W)$, with an L_{l2} structure [8–13]. In particular, the wear of PCBN FSW tool was investigated experimentally and numerically by Almoussavi et al., who determined that the tool shoulder periphery and probe side bottom are the most vulnerable parts suffering from wear issues [14]. As confirmed by Soresen et al., PCBN tools provide high quality finish when used for FSW of both ferrous alloys [15].

The study of weld process parameters, dissimilar to the Al/steel joint, is of great importance as it determines the heat input [16–19]. Elrefaey et al. [20] welded a plate of commercially pure aluminum (AA 1100 H24) to a plate of low-carbon steel (SPCC) in lap joint configuration and found that by increasing the rotation speed and decreasing the travel speed, grains of both aluminum and steel, in all characteristic areas, were coarsened and that a slight difference in pin depth (0.1 mm) has a significant effect on the performance of the lap joints. Kimapong and Watanabe [21] found that when welding A5083 aluminum alloy and a SS400 mild steel joint, it happens that increasing the tool rotational speed decreased the shear load of the joint because the higher rotational speed formed a thick FeAl3 intermetallic compound at the interface between aluminum and steel, and increasing the traverse speed of the tool increased the shear load of joints. Chen et al. [22] studied the effects of tool positioning on microstructures formed in the Al-to-steel interface region and reported that when the pin was close to the bottom steel piece, Al-to-steel reaction occurred, resulting in intermetallic outbursts formed along the interface, while when the pin approached the steel, a thin and continued interface intermetallic layer was formed. Wan et al. [23] enhanced the strength at the interface and eliminated the hook effect by adopting an enlarged pin head with circumferential notches.

In this paper, Friction Stir Welding between AA5083 H111 aluminum alloy and S355J2 grade DH36 structural steel dissimilar in lap joint configuration, with a polycrystalline cubic boron nitride with tungsten and rhenium additives tool, was presented. Sheets were welded under different conditions, FSW tool axial force, rotation speed, and welding speed. The quality of the FSW joint was evaluated by visual examination, X-ray tests (by ascertain the presence of macroscopic defects, such as incomplete welding penetration, cracks), and tensile and fatigue tests.

2. Materials and Methods

2.1. Materials

The used materials were AA5083 H111 aluminum alloy and S355J2 grade DH36 structural steel. The chemical composition and the mechanical characteristics of both materials are shown in Tables 1 and 2.

Table 1. Average composition (weight %).

AA 5083 H111						
Mg	Si	Fe	Cu	Mn	Cr	Zn
4.0 ÷ 4.9	0.40	0.40	0.10	0.40–1.0	0.05–0.25	0.25

S355J2 grade DH36												
C max	Si max	S max	Al	Mn	Cr max	Ni max	P max	Ti max	Cu max	Mo max	Nb	V
0.18	0.50	0.035	0.015	0.9–1.60	0.20	0.40	0.035	0.02	0.35	0.08	0.03	0.08

Table 2. Mechanical property.

5083 H111		
Rm (MPa)	$Rp_{0.2}$ (MPa)	A (%)
275	125	16
S355J2 grade DH36		
Rm (MPa)	Re min (MPa)	A (%)
490 ÷ 630	350	21

2.2. Methods

$500 \times 200 \times 4$ mm^3 thick plates were used in lap joint configuration. The steel plate was 1.5 mm thick while the aluminum plate was 4 mm thick. The aluminum plate was machined in order to remove a surface skin depth and thus favor the overlap on the steel, as shown in Figure 1a. The schematic set-up is shown in Figure 1b.

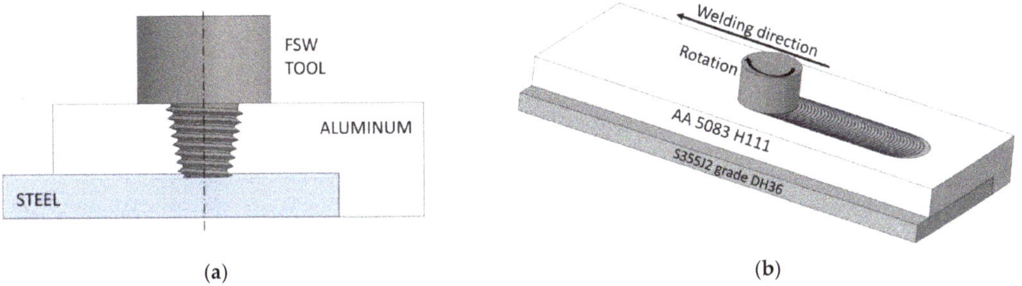

Figure 1. Schematic setup of the process in cross sectional (**a**) and axonometry (**b**) view.

Tests were carried out with a H. Loitz-Robotik (Hamburg, Germany) machine with a nominal torque of 70 Nm up to 3050 rpm, a tilt angle range of $+/-3°$, and a maximum vertical force of 60 kN (Figure 2). In particular, the present paper focuses on specimens produced with 1° tilt angle.

Figure 2. H. Loitz-Robotik Machine set up.

The tool (Figure 3), produced by MegaStir (Provo, UR, USA), with the abbreviation Q70 (70% PCBN, 30% WRe), was made of polycrystalline cubic boron nitride (PCBN) with tungsten and rhenium additives (WRe), which guarantee high thermal and mechanical properties to cope with wear problems. In addition, the tool has a micro hardness of 2600–3500 HV and a low coefficient of friction which affects the final roughness of the joint, improving the appearance [15].

Tool pin length was equal to 4.78 mm and shoulder diameter was equal to 36.8 mm. A profilometer was used to monitor the tool wear state (Figure 3).

Figure 3. Megastir Q70 (PCBN/WRe) tool.

Joints were made by using axial force control, where the tool penetration depth varies to keep the applied tool force constant. In addition, for each weld, a set of process parameters, which include axial force, rotational speed (ω), and welding speed (v), was identified (Table 3). For all trials selected, a tool tilt angle of $1°$ was adopted. Since a specific set of specimens was selected from an original batch of more than thousand trials, the numbering was kept as the original one and only the specimens assessed in the present investigation were considered.

To verify the presence of defects in the specimens, X-ray examination was carried out. Characterization of mechanical properties was carried out by testing lap shear specimens in the transverse direction, 25 mm width, using a Zwick (Ulm, Germany) Z600 machine with a 600 kN load cell at 2 mm/min at room temperature. The fatigue tests were performed on the same specimen geometry used for testing the static properties, with a Zwick (Ulm, Germany) Vibrophore 250 machine.

Table 3. Process parameters.

Samples (Original Batch)	Force (kN)	Travel Speed (mm/s)	Rotational Speed (rpm)	Weld Length (mm)
953	20.75	1.3	800	230
954	21.25	1.3	800	230
957	21	1.7	800	230
958	21	1.7	1000	230
959	21	1.7	800	430
987	21	1.2	1000	430
988	21	1.9	900	430
990	21	2.1	700	430
992	21	1.2	900	430
993	24.5	1.2	1000	430
994	28.5	1.5	700	430
998	24.5	1.5	800	160
995	26.3	1.7	700	430
984	21	1.7	800	430
985	21	1.7	800	430
999	21	1.7	800	150
996	30	1.9	700	275
997	26.3	1.9	700	150
1000	27	1.9	800	145

3. Results

Figures 4–6 show the top view of welds 994, 995, and 999, respectively. All welds presented a regular bead or reduced flash and a slightly rough surface. Despite that the process parameters adopted were different, a similar quality of beads can be explained considering that the combinations of parameters selected lay within the weldability window. Sample 994 was produced with a significantly larger force compared with sample 999, but the lower rotational speed led to less energy input to foster the plasticization. This is due to the tool material, since the polycrystalline cubic boron nitride tools produces smooth surfaces on the weld due to the low coefficient of friction between PCBN and metals [15]. Figure 7 shows a detail of the macro- and micro-section of sample 987.

Figure 4. Sample 994.

Figure 5. Sample 995.

Figure 6. Sample 999.

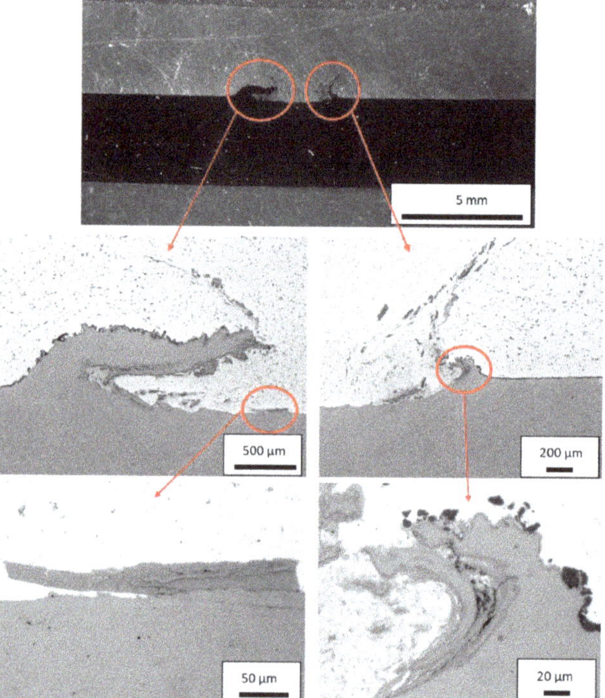

Figure 7. Sample 987 macro- and micro-section.

The picture shows the joint metallurgy within the stir zone (SZ), where refined and equiaxed grains were found because of the dynamic recrystallization linked to plastic deformation and heat input.

It is possible to observe the interconnection at the interface between Al and steel where the latter penetrates into the aluminum matrix due to the stirring effect performed by the pin. At the same time, a thin intermetallic layer at the interface between aluminum and steel was formed. It is possible to observe the action area of the pin and the characteristic hook shape generated by the roto-translational movement of the tool. In fact, in the interface area, the mechanical anchoring that the steel generated in the aluminum plate following plasticization and dynamic recrystallization is highlighted; the combination of mechanical anchoring together with diffusive phenomena between aluminum and steel are the two factors on which the mechanical strength that characterizes the joint directly depends.

Although the chemistry analysis is not in the scope of the present research, results reported in the literature for similar morphologies revealed the presence of intermetallics